The Science of Animal Growth and Meat Technology

The Science of Animal Growth and Meat Technology

Second Edition

Steven M. Lonergan
David G. Topel
Dennis N. Marple
Iowa State University

Academic Press is an imprint of Elsevier
125 London Wall, London EC2Y 5AS, United Kingdom
525 B Street, Suite 1650, San Diego, CA 92101, United States
50 Hampshire Street, 5th Floor, Cambridge, MA 02139, United States
The Boulevard, Langford Lane, Kidlington, Oxford OX5 1GB, United Kingdom

Library of Congress Cataloging-in-Publication Data
A catalog record for this book is available from the Library of Congress

British Library Cataloguing-in-Publication Data
A catalogue record for this book is available from the British Library

ISBN 978-0-12-815277-5

For information on all Academic Press publications
visit our website at https://www.elsevier.com/books-and-journals

Publisher: Andre Gerhard Wolff
Acquisition Editor: Patricia Osborn
Editorial Project Manager: Karen R. Miller
Production Project Manager: Omer Mukthar
Cover Designer: Christian Bilbow

Typeset by SPi Global, India

Contents

About the Authors

Contributing author Dr. James S. Dickson is a Professor in the Department of Animal Science and the Inter-Departmental Program in Microbiology at Iowa State University. Dr. Dickson's research focuses on the control of bacteria of public health significance in foods of animal origin. Prior to his appointment at Iowa State University in 1993, he was employed by USDA-ARS as a Research Food Technologist and lead scientist of the Meat Safety Assurance Program, at the Roman L. Hruska U.S. Meat Animal Research Center, Clay Center, NE, and was employed in the food industry for 3 years before joining USDA-ARS. He is a Fellow in the American Academy of Microbiology, is a Past President of the International Association for Food Protection, and is active in the American Society for Microbiology and the Institute of Food Technologists.

Author Dr. Steven M. Lonergan received his PhD in Animal Science with a minor in Biochemistry from the University of Nebraska-Lincoln in 1995. He has been on the faculty at Iowa State University since 1998. Dr. Lonergan's research focus at Iowa State University has been centered on discovery and thorough description of molecular factors that influence muscle growth and meat quality. He has demonstrated excellence in these endeavors as a researcher, teacher, mentor, and advocate for student achievement. Dr. Lonergan has been recognized by the American Meat Science Association and the American Society of Animal Science for Outstanding Teaching and Research. He is a Fellow in the American Meat Science Association.

Author Dr. Dennis N. Marple, (Professor Emeritus) received his BS (1967) and MS degrees (1968) from Iowa State University, PhD in Animal Science from Purdue University (1971), and conducted post-doctoral research at the University of Wisconsin, Madison (1971–73). He did research on animal growth and physiology, taught courses in animal growth and development and advanced meat science at Auburn University (1973–89), served as Head of the Auburn University Animal and Dairy Science department (1989–92) and as Head of the Animal Science department at Iowa State University (1992–2001). He is a Fellow and past president of the American Society of Animal Science.

Contributing author Dr. Joe Sebranek is Charles F. Curtiss Distinguished Professor in Agriculture and Life Sciences at Iowa State University where he holds the Morrison Chair in Meat Science. Dr. Sebranek received a BS in Animal Science from the University of Wisconsin—Platteville in 1970, and MS and PhD degrees in Meat and Animal Science and Food Science from the University of Wisconsin—Madison (1971 and 1974, respectively). He joined the faculty at Iowa State University in 1975 and has received numerous awards for teaching and research in meat science, including induction into the Meat Industry Hall of Fame in 2016.

Author Dr. David G. Topel, Professor and Dean Emeritus of the College of Agriculture, Iowa State University received the BS from the University of Wisconsin, Madison (1960), the MS in Meat Science from Kansas State University (1962) and the PhD from Michigan State University (1965). He was among the first to teach a course in meat animal growth and development while conducting research on pork quality at Iowa State University (1965–79). He served as Head of the Animal and Dairy Science Department at Auburn University (1979–88), served as Dean of the College of Agriculture at Iowa State University (1988–2000), and was later named to the M.E. Ensminger Endowed Chair of Animal Science.

Preface

This book was developed to help provide an understanding of the principles of meat science and technology starting with prenatal growth of domestic animals through postnatal growth. It was prepared for students with an animal science interest as well as those actively involved in animal production and meat processing. The book relates the science of animal production to technologies and meat quality traits that are important in the meat and animal industries. It provides the student and individuals with a unique opportunity to associate animal growth traits, production, and marketing traits to carcass quality, meat tenderness, meat color, and meat processing characteristics. Learning how to solve problems and thinking independently are important concepts that an individual can use to be successful when working in the agricultural industries that includes farming. This book provides concepts that can be helpful for independent problem solving associated with the animal and meat industries.

The animal and meat industries had a colorful and sometimes very difficult economic history. Lessons were learned from the historical years that resulted in the modern and high-technology meat and animal industries we have today. Therefore the first chapter provides a short introduction of the history of the industry. The subsequent chapters provide principles of animal growth and development to carcass composition and meat quality traits. Other chapters provide information on the slaughter process of animals, muscle structure and meat tenderness, meat quality, and meat safety and microbiology. The concluding chapters are on meat processing and packaging technology.

A large percentage of animal science students are interested in preveterinary medicine and many are from an urban background. This book will provide these students with concepts and principles that will give them a good background for understanding the information on animal agriculture presented in the advanced animal science courses.

The authors have used many colorful illustrations to emphasize important relationships between animal growth and carcass traits, meat quality, and processing characteristics.

Acknowledgments

We are grateful to the many individuals who provided special help for the preparation of this book. Professor James Dickson, Animal Science Department, Iowa State University (ISU), was very helpful in the preparation of the meat microbiology and safety chapter. Professor Dickson and Robert Hubert provided most of the figures for Chapter 12. Also, Professor Joe Sebranek, Animal Science Department, ISU, prepared Chapter 15 to fill a void regarding packaging technologies used in the preservation, shipment, and merchandizing of meat products. We want to pay tribute to Professor Joe Cordray, Animal Science Department, ISU, for his help with the meat processing chapters and to Bob Elbert, Media Production Specialists, ISU, for the photos used for the figures in Chapters 13 and 14. Important contributions were provided by Professor Emeritus Gene Rouse, Animal Science Department, ISU, for data and figures provided for the chapter on growth curves and growth patterns, and also for figures in the chapter on methods to measure body composition of domestic animals A special thanks to Professor Emeritus Dean Henderson, University of Wisconsin, River Falls, for his contributions in reviewing Chapters 13 and 14 and providing figures for Chapter 1. Roslyn Punt, Animal Science Department, ISU, was responsible for preparing many of the growth curves for the figures in Chapter 6. Also, Marley Dobyns, Animal Science Department, ISU, prepared figures for the muscle structure and function chapter. The contributions of Roslyn and Marley were significant in the development of these chapters. This was a significant contribution. Finally, this book would not have been possible without support of Professor Maynard Hogberg, former Chair, Animal Science Department, ISU. His help is greatly appreciated.

CHAPTER

1 Historical perspectives of the meat and animal industry and their relationship to animal growth, body composition, and meat technology

INTRODUCTION

There are significant relationships between the regulation of animal growth and body composition and meat quality traits of domestic animals. Animal growth traits and carcass characteristics have great influences on the value of the live animal for both breeding value and retail meat value. Therefore it is important to understand growth and development concepts when management decisions are made for livestock production systems.

This book will provide fundamental science-based concepts as well as applied and practical concepts from prenatal growth to postnatal growth of cattle, sheep, and pigs. This book is unique, as information is also presented that relates growth and development traits to the carcass value, meat retail characteristics, meat processing, and meat storage traits that are important at the wholesale and retail markets.

HISTORICAL PERSPECTIVES FOR THE ORIGIN OF MEAT-PRODUCING ANIMALS AND THEIR GROWTH TRAITS

DOMESTICATION OF CATTLE

The origin of cattle is often traced to the Indian subcontinent, as skeletal remains have been recovered in this region of the world. The remains are representative of the Zebu cattle. The skeletons of the male cattle were large (6 ft. at the withers) and their horns measured 3 ft. Their forequarters were large and their hindquarters were less developed. Their growth potential was large. Domesticated cattle that dated back to 6500 BC were also unearthed in Turkey.

When cattle were domesticated, major growth and conformation changes took place depending on the use of the cattle by the people who did the domestication. Cattle used for draft purposes were selected for large size, and cattle selected for milk

The Science of Animal Growth and Meat Technology. https://doi.org/10.1016/B978-0-12-815277-5.00001-9

and meat production were smaller. Cattle became important to the culture of families and they were often drawn on the walls of tombs, reflecting the importance to the family lifestyle. An example is shown in Fig. 1.1. This figure is from an Egyptian tomb. During the domestication process, cattle were selected to meet the needs of the people in the different regions of the world, and this resulted in a large number of cattle with different conformation and growth rates. In more recent years (1725–95), Robert Bakewell from England used inbreeding techniques to develop breeds of cattle. He was best known for the development of the Durham breed. It became a very popular cattle breed in England and Europe in the 1800s. When the Durham cattle were exported to the United States, they became known as the Shorthorn Breed (Fig. 1.2). It was split into a Dairy Shorthorn and Beef Shorthorn for milk and meat production, respectively.

The 19th century was a significant time period for cattle breed development. Many of our current breeds were developed in the 19th century. Some examples are the Angus, Hereford, Holstein, Guernsey, and Brown Swiss. The growth and development patterns of these breeds have changed extensively from the 19th century. These changes will be described in the growth and development chapters of the book.

DOMESTICATION OF THE PIG

Historical information on the domestication of the pig indicated that the pig was separately domesticated in several distinct areas of the world more than 8000 years ago. Some pigs were domesticated in the Alps region of Europe and the Baltic Sea area. There is also much evidence that the pig was domesticated in China. The conformation and growth patterns of the Chinese pigs were very different from the domesticated European pig (Fig. 1.3). The genetic divergence between the European

FIG. 1.1

An example of cattle from the walls of an Egyptian tomb.

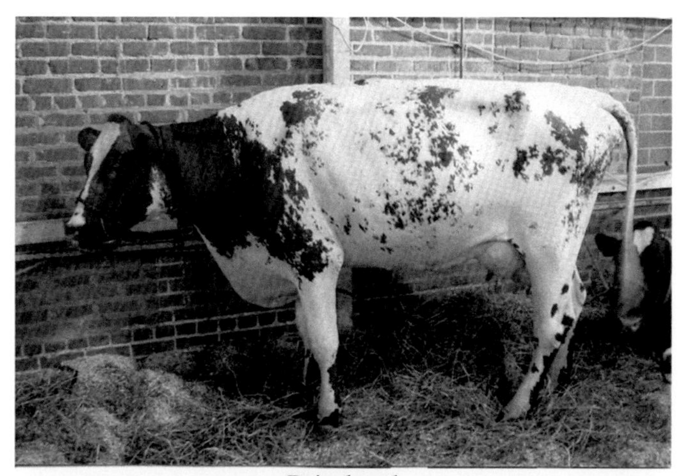

Dairy breed

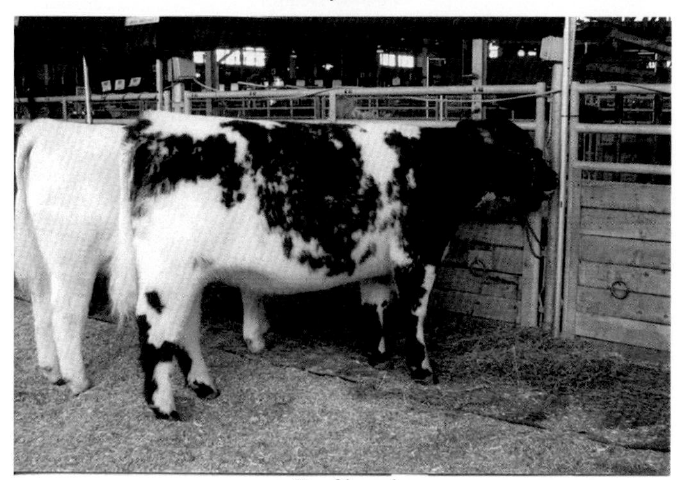

Beef breed

FIG. 1.2

Examples of the Shorthorn Breed developed by Robert Bakewell from England.

and Chinese pigs is estimated to be 500,000 years ago. As the pig was domesticated, more swine breeds started to develop. Darwin, in 1871, reported two species of domestic pigs that obtained their origin from the wild boar in Europe and Asia. Recent information from molecular genetic research indicates that Asian pigs were introduced to the European population in the 18th and 19th centuries. Large differences in litter size and growth rate existed in these pigs. The Chinese pigs had very large litters compared to the European pigs, and the European pigs had much faster growth rates. Both the European and Asian pigs had a high percentage of fat compared to muscle. The people in this time period preferred fatter meat and meat products and therefore selected for fatter pigs.

Chinese example

European example

FIG. 1.3

An example of the Chinese pig selected to meet the needs for the Chinese population during domestication, and an example of the pigs that resulted from the domestication process in Europe.

The majority of the US pig population obtained their origin from Europe, and most of the current US swine breeds were imported from Europe. In the 19th century and the early part of the 20th century in the United States, swine breeds were also selected for more fat than muscle, as pork fat was a stable part of the human diet in the United States. Also, the fat provided a lot of good flavor to the meat and processed meat products, and this was preferred by the consumer. Fig. 1.4 gives an example of the pigs in the United States in the first half of the 20th century. In the last half of the 20th century, major changes in selection occurred in the body composition of

FIG. 1.4

An example of a pig from the first half of the 20th century.

the pig population. Fat was greatly reduced and muscle was increased by intensive genetic selection. The rapid change in body composition of the pig population in the 1960s to the 1990s resulted in adjustment problems for most of the very muscular pigs. They became very susceptible to stress conditions. This stress impact, often associated with a very rapid increase in muscle deposition during the growth process, caused significant financial losses to the swine industry, as the stress-susceptible pigs often died when transported to market. This is a very important topic in the historical development of the modern pig population and will be presented in more detail in several chapters.

In the 1990s, genetic tests were developed to eliminate the stress condition (porcine stress syndrome) from the pig population. The pig industry made great changes in the composition and production traits of the domesticated pig from the 1960s to present. More changes in body composition and growth rate were made during this period than in the previous 150 years. These changes will be presented in the growth and development section of this book.

DOMESTICATION OF SHEEP

It is thought that sheep were first domesticated in 9000 BC in what is now known as Iraq. Syria, Iran, and Turkey were also recognized as some of the first regions of the world to domesticate sheep. Sheep were popular for trading activities, and domesticated sheep rapidly moved from Iraq to Europe, Africa, and Asia. Sheep were genetically altered for growth and development traits to improve meat, milk, and wool production. Some sheep were developed into breeds with special traits for wool, milk, and meat production. There are many sheep breeds around the world. The most popular sheep breed in the United States is the Suffolk (Fig. 1.5). The Suffolk was genetically selected for great growth traits, resulting in the largest mature weight of the US breeds. It is also a very muscular breed. Growth and development patterns for sheep will be presented in the growth sections of the book and will be related to body composition.

FIG. 1.5

An example of an excellent and modern Suffolk.

Courtesy of Fisher Suffolks, West Des Moines, Iowa.

It is evident that animal growth and its relationship to carcass composition was an important topic for sheep producers to consider when sheep were domesticated. These same growth and development traits are altered today to design the "seed-stock" sheep population to meet the economic characteristics required for a profitable industry. The Suffolks shown in Fig. 1.5 are examples of genetic selection programs of sheep for profitable traits in the sheep industry. The genetic controls for animal growth will continue to be altered to reduce production costs and improve carcass composition and meat quality. Therefore a historical knowledge of the most important factors that regulate growth and development and body composition of domestic animals can be helpful when best management practices are used for production systems in the livestock industry.

HISTORICAL RELATIONSHIP BETWEEN MAN AND DOMESTIC ANIMALS

The drawings (Fig. 1.1) on the walls of caves and tombs dating back 4–5 thousand years ago indicate an important and positive relationship between man and domestic animals. This positive relationship still exists today for many people. The domestication of the dog probably took place in Eastern Asia about 15,000 or more years ago. This is based on recent DNA evidence. They were domesticated for companionship and often were part of the family thousands of years ago.

RELIGION AND ANIMALS

In Asian countries, oxen and water buffalo were used for transportation and field work and often they became part of the family. Many families would not eat meat from the water buffalo or oxen. Some domesticated animals became part of man's religion and also became part of their social culture. The Buddhist religion is an example of this. Many Buddhists are vegetarians and some are against the slaughter of animals for meat consumption. The Hindu religion considers cattle sacred and they are protected in India.

MEDICAL SCIENCE AND ANIMALS

The foundation of medical science was greatly influenced through the use of domesticated animals for medical experiments, and the longer life expectancy of man continues today because of the medical science database that was obtained from using domestic animals as experimental models. It is of interest that the physiological functions of pigs are very similar to humans. Therefore the pig is a very good model for human medical research. Fig. 1.6 shows the influence of vitamin B12 on the growth rate of pigs used as an experimental model for the impact of vitamin requirements. The pig on the left received no supplemental B12 and the pig on the right received B12 in the ration. Note the difference in size. Early studies used the pig as an experimental animal to determine B12 requirements for humans. Our society was built with the use of domesticated animals as a source of critical medicines. An example would be the hormone insulin, which can be extracted from the pancreas of the pig. When this procedure is used, the pancreas is harvested during the slaughter process. The extracted insulin was a "life saver" for people who had diabetes. Insulin obtained from animals was the major source until molecular genetic techniques were developed and bacteria were modified to produce insulin. Without domesticated animals, our society would not be as advanced socially, scientifically, or technically.

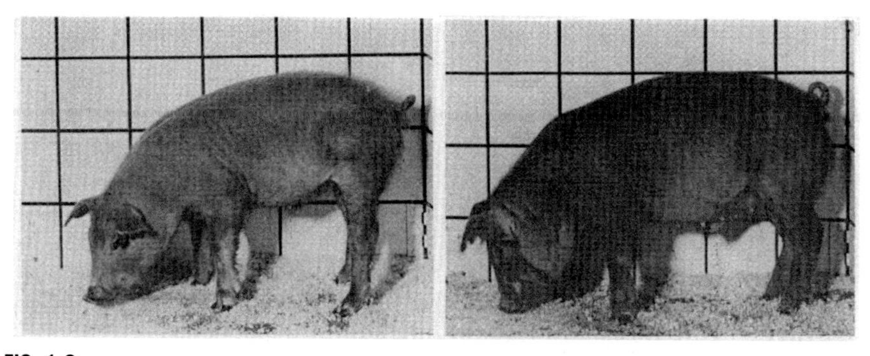

FIG. 1.6

An example of pigs in an experiment to study vitamin B12 on growth rate and general health conditions. The pig on the left received no supplemental source of vitamin B12, and the pig on the right received 5 μg of vitamin B12 per pound of ration.

From the Council of Agricultural Science and Technology (CAST), Ames, Iowa, Food From Animals, 1980. Courtesy of Vaughn C. Speer, Animal Science Department, Iowa State University, Ames, IA.

HISTORICAL RELATIONSHIP BETWEEN THE LIVESTOCK AND MEAT INDUSTRY

There is an interesting history relating the livestock and meat industry in the United States. They grew together and often conflicted with each other on prices paid by the packer to the livestock farmers and ranchers. Some ranchers and farmers worked together and built packing companies to compete with the traditional packer, but most of the farmer-owned packing companies were not very successful and many were purchased by the traditional packing companies or they were closed.

ORIGIN OF THE MEAT INDUSTRY IN THE UNITED STATES

In the United States, colonial farmers working together are given credit for becoming the first meat packers. The colonial farmers slaughtered pigs, cattle, and sheep at their farms; cut the carcasses into smaller portions; salted the meat cuts; and packed them into wooden barrels. The barrels were shipped to cities and the meat distributed to small meat markets. Therefore the name "meat packers" was established.

William Pynchon of Springfield, Massachusetts is often recognized as the first individual meat packer in the United States. Sam Wilson, from Troy, New York, was also an early pioneer in the meat industry, as he packed salted meat into barrels and sent the meat to the US army for the war of 1812. The barrels were stamped "U.S." and were often referred to by the soldiers as "Uncle Sam's beef" or "Uncle Sam's pork." The term rapidly caught on for everything related to the US government and this is how the term "Uncle Sam" was started; it is still used today as a name for some US government programs.

HISTORY OF THE OLD LINE PACKERS

With the establishment of the railroad system in the United States, especially east of the Mississippi River, livestock could be transported to major cities. Therefore packing companies were built near these cities, and some of the packing companies became very large. The largest center for stockyards and packing companies was in Chicago. The Union Stockyards in Chicago provided a place where ranchers and farmers would bring their cattle, hogs, and sheep to be held before they were sold to the packing companies. Beef cattle were delivered to the Union Stockyards from as far away as Texas, Oklahoma, and Kansas. Hogs and sheep from throughout the Midwestern states were marketed through the Union Stockyards. It was the center of livestock trading in the United States from the 1850s to the 1950s. During these 100 years, Armor, Swift, Wilson, Morris, and Cudahy packing companies controlled the Chicago meat markets. These companies expanded to other regions of the United States as well as other countries. At one time, these five companies controlled more than 80% of the meat-packing industry. These companies were founded by hard-working, aggressive, and demanding individuals. Philip Danforth Armor (Fig. 1.7) is an example. The first packing plant that Armor started in Chicago was a small wooden building (Fig. 1.8), and from this small wooden building, Mr. Armor built

FIG. 1.7

A photo of the pioneer meat packer Philip D. Armor.

Courtesy of the Animal Science Department, Iowa State University.

FIG. 1.8

A photo of the Armor packing plant first built in Chicago in 1875.

Courtesy of the Animal Science Department, Iowa State University.

FIG. 1.9

An example of a large Armor plant in Chicago in 1926.

Courtesy of the Animal Science Department, Iowa State University.

a giant company with packing plants located in many cities in the United States and other countries. Fig. 1.9 shows the Armor plant in Chicago in 1926. Gustavus Swift was also a significant pioneer in the meat industry. He was just as aggressive as Armor and established packing companies as large or larger than Armor. Swift developed the refrigerated rail road car for transportation of fresh and processed meat throughout the United States. This was a major contribution for the development of new markets for the meat industry and provided higher quality meat products at the retail and wholesale meat distribution centers. These packing companies were often located in the same city and by the same stockyards. Some of the major locations were Kansas City, Indianapolis, Sioux City, Omaha, New York, East St. Louis, Milwaukee, Spokane, Fort Worth, and Denver. They also had operations in Argentina, Brazil, and Uruguay.

SOCIAL CONDITIONS

The working conditions were difficult in the early packing plants and many of the workers were immigrants from Europe. The pay was low, and in 1904, the first big strike in the meat industry took place. More than 50,000 people walked out of the packing companies' plants. The strike stopped when better wages were obtained by Union leaders. The working conditions, however, were still poor, and a young man named Upton Sinclair published a book, "The Jungle," on the working conditions and sanitation in the packing plants, social issues with women and men working at the same plants, and lack of good inspection standards. The book stimulated increased government oversight and resulted in the passing of the Federal Meat Inspection Act of 1906. This was a very good act and lasted until the late 1960s, when it was altered to establish the Wholesome Meat Act to address the outbreaks of *Escherichia coli* and other microorganisms causing food poison problems from meat and meat products.

THE 1940-TO-1980S MEAT INDUSTRY

Some of the outstanding meat companies today started out as small companies in the 1920s to 1940s. The Oscar Mayer Company started as a quality sausage company in Chicago and expanded to several cities, including moving its headquarters to Madison, Wisconsin. In the 1950s, Oscar Mayer management established an outstanding cooperative program with the Meat Science Faculty at the University of Wisconsin. The University of Wisconsin Meat Science Program was coordinated by Dr. Robert Bray (Fig. 1.10), an early PhD and pioneer in the Meat Science field in the United States. Cooperative research, education programs, and student internships were established. This arrangement strengthened the science base for the meat industry, and the Oscar Mayer Company became a leader in the industry for quality control, marketing, packaging materials, and quality processed meat items.

The Oscar Mayer Company used unique and very successful methods to promote their sausage and processed meat products. The Oscar Mayer Wiener-Mobile (Fig. 1.11) is an example. This unique promotion concept was successful throughout the United States. Often students were hired to drive the Oscar Mayer Wiener-Mobile and to do promotions with special songs for children.

Other meat companies in the Midwest also expanded in the 1940s to 1980s. Geo. A. Hormel & Co. (now Hormel Foods Corporation) in Austin, Minnesota, is an

FIG. 1.10

A photo of Dr. Robert Bray, outstanding educator in the meat science field.

Courtesy of the Department of Animal Sciences, University of Wisconsin, Madison.

FIG. 1.11

A photo of the Oscar Mayer Wiener-Mobile used for marketing of Oscar wieners and other processed meat items.

example. Hormel Foods developed new technology for canning meat and marketed the semisterile canned ham. As a historical note, Hormel Foods came out with their first canned ham in 1926 and introduced the SPAM brand in 1937. The SPAM product has ham-like flavors, and there is some actual ham in the product. This unique canned product is stable at room temperatures. The product is marketed under the SPAM brand (Fig. 1.12). SPAM luncheon meat became a popular item for the soldiers in World War II and continued to be a popular meat product when

FIG. 1.12

An example of SPAM Products marketed at the retail level.

they returned to the United States after the war. This canned, sterile product was low in price and had very good quality traits; it is still a popular product in today's meat markets.

The John Morrell Company in Ottumwa, Iowa, also became a major pork and beef packer in the Midwest that purchased livestock directly from farmers. They competed with other companies such as the Rath Packing Company in Waterloo, Iowa, and Hormel Foods in Minnesota to purchase livestock directly from the farmer, rather than through a central market such as the Union Stockyards in Chicago. The direct purchase of livestock from farmers resulted in better prices for the farmers. The packing companies that purchased livestock directly from farmers also provided educational programs for the farmers to improve production and management of the livestock. Often, these programs were in cooperation with professors and the county extension personnel at the Land-Grant Universities. These cooperative programs provided a good working relationship between the meat industry personnel and faculty and extension staff at the agricultural universities in the United States. These cooperative activities continue today.

The companies in the meat and food industry hire many students from the agricultural universities in the United States and sponsor research projects to strengthen the industry. As companies like Hormel Foods and Oscar Mayer became larger and stronger, the packing companies associated with the "Central Markets," such as the Chicago Union Stockyards, started to decline. Their facilities were getting old and the profits were poor. As a result, the Stockyards and large companies, such as Armor and Swift, started to close their old plants in the late 1950s and early 1960s, resulting in the closing of many stockyards associated with the old packing plants. Also at this time, transportation of the meat and meat products changed from refrigerated railroad cars to refrigerated trucks. These factors influenced the meat industry to consider the construction of packing plants where the livestock were raised. This concept opened opportunities for a new direction for the meat and livestock industries.

OPPORTUNITIES FOR RURAL AMERICA AND THE DEVELOPMENT OF THE MODERN MEAT INDUSTRY

The next major historical development for the meat industry started in 1957, when a modern pioneer of the meat industry, Mr. Andy Anderson, established a major, one-story meat-packing plant in a small rural community, Denison, Iowa. He selected Denison because the workforce needed jobs; they were a hardworking and dedicated workforce and there were no other packing companies in the area. Most important, however, the new pork plant was located in one of the most populated areas in the United States for livestock production. The new company could purchase all of the pigs needed for the capacity of the slaughter plant within several miles of the Denison plant. This was good for the farmers and good for the new pork processing company. They could purchase the livestock they needed directly from local farmers. Another significant change was the concept for a single-species plant compared to the older multispecies plants that had three to four floors and slaughtered cattle, sheep, and pigs. The new plants had only one floor and were much more efficient than the older plants. This new pork packing company was very successful.

The next project for Mr. Anderson was to establish a beef slaughter plant based on the same concepts as the Denison pork plant. Mr. Currier Holman joined Mr. Anderson as a major partner and they established, in 1960, the Iowa Beef Packers (IBP) Company. These two men introduced new, efficient, and important quality concepts to the packing industry as they built IBP to become one of the leading packing companies in the United States for cattle and pigs.

THE BOXED BEEF CONCEPT

One of the most significant contributions from IBP was a major change on how beef carcasses were sold and distributed to the retail market. Rather than shipping the whole carcass or half of a carcass, IBP cut the carcass into smaller cuts and placed them into a box and sold the box of beef to the retailer. This greatly reduced shipping costs and significantly improved efficiency for selling beef at the retail markets. Selling beef in a box sounds like a simple idea, but it was a difficult marketing concept as the very strong Meat Cutters Unions in the large cities prevented the boxes of beef from being delivered to the retail markets. They were concerned about a great reduction in jobs if the boxed beef concept was accepted. After some difficult financial times, the IBP management finally got through Union controls and boxed beef was accepted as a new marketing concept for the industry. The boxed beef marketing style, along with many other new concepts for improving the efficiency and quality of the beef packing industry, allowed IBP to become one of the largest beef and pork processors in the United States in the 1980s and early 1990s.

THE CHANGE TO ADDED-VALUE PRODUCTS

In the 1990s, many of the traditional packing companies expanded from commodity products, such as boxed beef and pork, to value-added meat products; for example, ready-to-consume roasts, chops, deli products, taco meat, pizza toppings, and hors d'oeuvres. IBP, Hormel Foods, Oscar Mayer, and Cargill are examples of the traditional packing companies that expanded into premium, added-value food products. Some of the expansion was by the purchase of established food companies that already had outstanding markets. This was another major change in the historical development of the meat industry. The new added-value concept was expanded to "ready-to-serve" meat and food products that included complete dinners that could be heated in a microwave oven for a very convenient way to serve a family meal (Fig. 1.13).

FIG. 1.13

An example of added-value meat products.

BRANDED FRESH BEEF PRODUCTS

Branded beef concepts were started in the 1970s by purebred beef cattle associations to promote beef at retail stores and high-quality restaurants in the United States. The most successful of the branded beef programs was Certified Angus Beef (Fig. 1.14). This program was established and supported by the cattlemen producing Angus cattle. The program established a new concept for marketing high-quality beef in the United States and in international markets by promoting Angus beef with a high degree of marbling. This program resulted in another important marketing achievement in the historical development of the meat industry as it was started and is still under the management of cattlemen associated with the Angus breed. In the 1990s, certified Angus beef became very popular. In order to meet the great demand for Angus beef, the American Angus Association needed to establish new concepts for selection of Angus cattle that had a high degree of marbling (intramuscular fat) in the meat. The selection program included the use of growth and development and body composition principles that will be discussed in several chapters of this book.

FIG. 1.14

An example of certified Angus beef in the retail market.

Courtesy of the Certified Angus Beef brand.

MERGERS IN THE MEAT INDUSTRY

In the early years of the 21st century, major mergers started to occur between the red meat industry and the poultry industry. The largest poultry company, Tyson, purchased one of the largest beef and pork processors, IBP. All of the IBP companies became part of the Tyson Company. Tyson became the largest producer of meat and meat products in the United States. The Smithfield Company also expanded significantly in the 1990s and the early years of the 21st century with the purchase of other major meat processing and packing companies. Smithfield Food is ranked number five and Tyson Foods is ranked number one in sales. JBS-USA Holding Inc. is second, Cargill Meat Solutions is third, Sysco Corp. is fourth, Con Agra Foods Inc. is fifth, and Hormel Foods is ranked sixth.

The meat industry transformed from a low-technology industry in the early years of its existence to a quality industry in the 21st century through the application of good science and modern technology. Therefore the meat industry today is a supplier of quality protein foods to the world. The quality protein foods are a result of the livestock industry working with the meat industry to provide a nutritious meat source to consumers around the world. This book provides a science base of information that can be used in the livestock and meat industry and thereby result in the production of quality meat and meat products for the 21st century.

SUMMARY

The relationship between the genetic regulation of animal growth and development and body composition is high. These relationships have changed as pigs, cattle, and sheep were domesticated in different parts of the world, starting approximately

6000–10,000 years ago. Cattle were domesticated in the Indian subcontinent and pigs in China and Europe (the Alps region and the Baltic Sea region). It is thought that sheep were first domesticated in what is now known as Iraq in 9000 BC. As the animals were domesticated over the years, their growth and body composition patterns were altered to meet the needs for the individuals who were responsible for the domestication. That is why there is so much genetic diversity among domestic animals. Domestic animals have had special relationships with man for thousands of years. Some relationships are for companionship, some are for religions, such as the Buddhist and Hindu religions, and some were for field work and transportation. Domestic animals also contributed to the improvement of human health around the world. For example, the pig contributed to a source of insulin for people with diabetes.

There is an interesting historical relationship between the livestock industry and the meat industry. The meat industry started with farmers slaughtering and packing pork, beef, and lamb for the meat markets in the colonial years of US history. The next phase was the establishment of the large industrial companies, such as Armor, Swift, and Morris, that were associated with the Stockyards for a source of animals. The next phase was a movement of the meat companies to the source of the animals, rural America.

The current structure of the meat industry has resulted in an "added-value" industry by mergers with food companies and the development of new, high technology produced meat products. New marketing concepts were promoted with much success, such as the Angus Certified Beef Program. Over the last 200 years, the meat industry has transformed from a low-technology industry to a high-technology supplier of quality protein foods for the US consumer and for consumers around the world.

QUESTIONS FOR STUDY AND DISCUSSION TOPICS

1. Who was the first individual meat packer in the United States?
2. How did domestication of animals influence the growth and development pattern of cattle, pigs, and sheep?
3. Give an example of the importance between the domestication of animals and their impact on human health?
4. Who was Robert Bakewell and what did he contribute to the early development of beef cattle breeds?
5. When was the pig domesticated and in what regions of the world?
6. Describe the marketing concepts available to farmers for selling live cattle and pigs in the Midwest between 1850 and the 1950s.
7. Describe the most significant impacts the Iowa Beef Processors (IBP) had on the beef industry.
8. Provide an example of a successful marketing program for selling "branded" fresh beef.
9. Compare the meat industry in the 1850s to the meat industry in the 21st century.

CHAPTER

Prenatal growth and its relationship to carcass and meat quality traits

2

INTRODUCTION

The prenatal period is a time of rapid relative growth and is a time during which the organism doubles in size at an ever decreasing rate. This is a period of remarkable transformation from a fertilized ovum to an organism having fully defined muscles, organs, and nervous system at birth. Although more maturation of these structures will take place during postnatal growth, the key events of organization and differentiation occur during the prenatal phase of growth. Prenatal growth may be divided into three periods: ovum, embryonic, and fetal to help organize the discussion of prenatal events.

OVUM PERIOD

The ovum period starts with fertilization and restoration of the complete number of chromosomes, with one chromosome of each pair of chromosomes being derived from each parent. Initially, cell division occurs within the confines of the zona pellucida, the membrane originally covering the ovum and penetrated by the sperm cell. Cell division continues in this confined space through the morula and early blastocyst stages, so there is a corresponding decrease in cell size or compaction of cells during the ovum period (Fig. 2.1).

The morula stage is defined as the stage of development during which the dividing cell mass has 20–30 cells and resembles a mulberry, hence the term morula. It is at the morula stage that fertilized ova are most suitable for transfer to surrogate females, a practice widely used in the dairy and beef industries to expand the genetic impact of superior bulls and cows. The process is termed embryo transfer although the developing organism has yet to reach the embryonic stage of development, and it is done successfully in other species as well. Although cattle typically produce only one ovum at a time, they can be induced to produce 10–20 ova at once by treatment with reproductive hormones. After breeding, the resulting fertilized ova are allowed to develop naturally to the morula stage (d 7 of development) and are then transferred to recipient females who have had their reproductive cycles synchronized with that of the donor female. In the event that too many fertilized ova have been collected for immediate transfer, the excess ova may be frozen and later transferred to recipient females. The process has been done without surgical intervention in cattle for many years and similar procedures are being developed for other species. However, the

The Science of Animal Growth and Meat Technology. https://doi.org/10.1016/B978-0-12-815277-5.00002-0

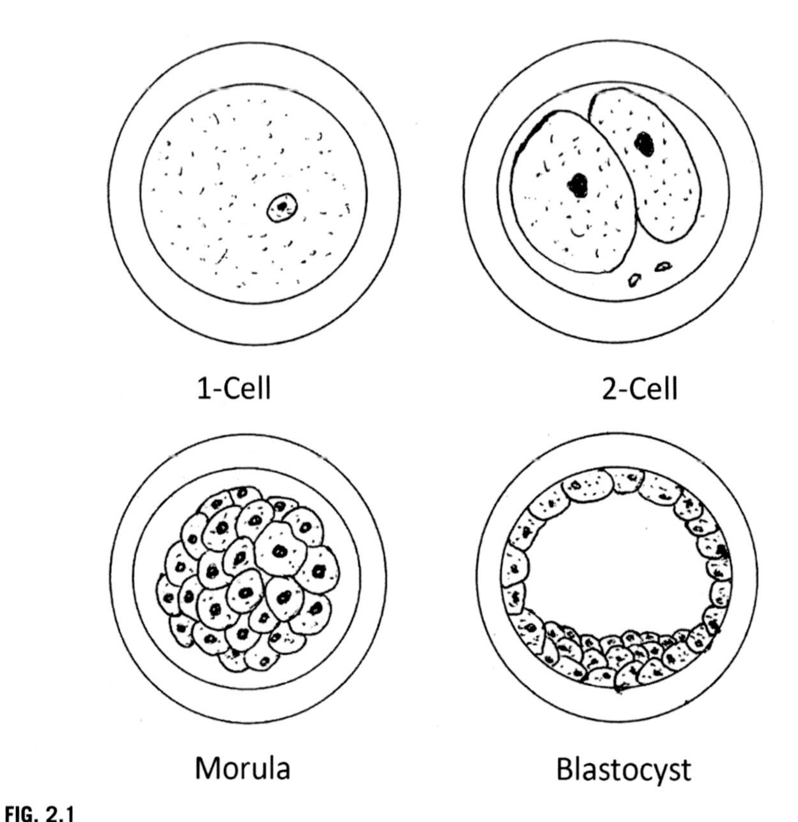

FIG. 2.1

Cellular organization during early embryonic development.

technology to freeze and thereby preserve fertilized ova varies among species in that the technique is more successful for cattle than for other farm animals.

The ovum period is a period of hyperplasia (an increase in cell number) and is also a time of morphogenesis, the process whereby cells begin to become compacted within the morula to form blastomeres or cells that will develop into specific tissues depending on their location within the blastocyst, a hollow, ball-shaped structure with cells lining the zona pellucida. The inner cell mass will become the embryo, and the outer cells lining the zona pellucida are the trophoblast which will become the embryonic membranes, including the placenta, amnion, and chorion. Ultimately, the zona pellucida ruptures, resulting in hatching of the blastocyst which then takes on a more elongated, filamentous shape composed of the tissue disk of the inner cell mass and the developing membranes. This marks the end of the ovum period and start of the embryonic period.

EMBRYONIC PERIOD

The embryonic stage of development is characterized by rapid cell division, including the processes of cell determination and differentiation. Cell determination is defined as when cells become committed to a specific developmental pathway but are not yet cells of that particular type. These cells destined to become particular tissues migrate to specific regions of the embryo where they undergo further transformation and become differentiated into primitive tissues such as muscle, connective tissue, or organs. Differentiation is the transformation of a precursor cell into a specific cell type. By approximately one-third of the gestation period, the embryo has undergone a dramatic transformation from undifferentiated cells to an organism distinctly resembling the fetus at birth.

There is limited increase in cell size or hypertrophy during the embryonic period, and the first noticeable structure to form is the primitive streak, a thin line in the tissue disk of the hatching blastocyst creating the longitudinal axis along which the bilateral symmetry of the organism is defined. The innermost cell layer of the inner cell mass undergoes cell division and ultimately forms three distinct germ layers, the endoderm, mesoderm, and ectoderm, and the tissues of the body will form from each layer. The three cell layers in this elongated disk proliferate and fold inward, creating a cylindrical structure that is thicker on one side. This process is termed gastrulation.

The next distinct structure formed is the notochord, a thick line of cells parallel and interior to the primitive groove. The notochord ultimately is the structure along which the axial or central skeleton is formed. Cells of the ectoderm adjacent to the notochord thicken along each side of the central axis, forming the neural groove, and ultimately fold together, forming the neural tube which becomes the spinal cord and the brain. Other cells of the ectoderm form the sensory nervous system; mammary glands; sweat glands; and epidermal layer of the skin, hair, and hooves.

The middle layer of cells continues to evolve into the somites, buds of tissue seen opposite each other along the axis of the developing embryo (Fig. 2.2). The pairs of somites form first near the developing head. The last pair to form is toward the tail end of the embryo. The number of pairs of somites present in the developing embryo is an accurate indicator of the age of the embryo. The somites differentiate into sclerotomes, which form the skeleton; myotomes, which form the skeletal muscles; and dermatomes, which form the dermis of the skin. Sclerotomes are located adjacent to the neural tube and in the medial portion of the somites. These cells produce the axial or central skeleton of the embryo. Myotomes evolve from the dermatomes in that the outermost (lateral) cells of the dermatomes become the dermis and the inner cells become the myotomes and ultimately the muscles of the body.

Limbs of the body develop from limb buds, somatic cells that become specialized in their structure and control to the extent that precartilage mesenchyme cells begin to accumulate under the dermis and adjacent to where the limb should form. Cells in the ectoderm thicken and form a point to which mesodermal cells migrate and growth factors are produced. The skeleton, connective tissues, and blood vessels of the limb form from lateral plate mesodermal cells, and myotome cells from somite mesoderm form the muscles. Development of the limb occurs in a proximal to distal manner, with the most immature portion of the limb at the end of the developing limb. The study of cellular organization and differentiation associated with limb formation has received intensive study, and many detailed mechanisms related to the expression of growth factors regulating these developing tissues continue to be described.

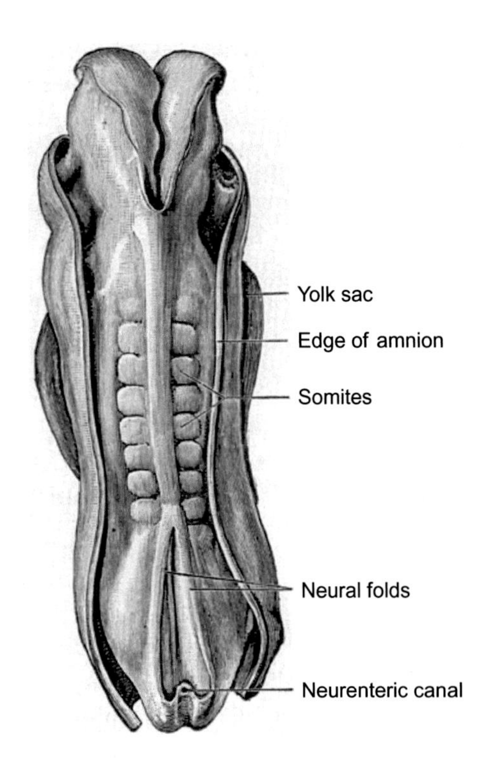

FIG. 2.2

Early development of a human embryo showing the appearance of pairs of somites.

The mesoderm is also the origin of tissues of the circulatory system, including the heart, arteries, capillaries, veins, blood, lymphatic vessels, and lymph. In addition, the mesoderm gives rise to some tissues of the digestive system, respiratory system, and urinary system, including the kidney, ureter, and urethra.

The inner layer of cells of the inner cell mass forms the endoderm which becomes the digestive tract, including the esophagus, pharynx and base of the tongue, liver, pancreas, lungs, and bladder. The alimentary canal formed from the primitive gut is continuous with the exterior of the body and is lined with a layer of epithelial cells of various types.

MYOGENESIS

Skeletal muscles originate from mesodermal cells of the somites. The muscles of the back are from the epaxial myotome (near the neural tube) whereas muscles of the ribcage and limbs are formed from the hypaxial myotome (the area most distant from the neural tube). In each instance, muscles do not simply appear but are the result of a carefully orchestrated process involving the interaction of many growth factors and regulatory genes. The process whereby cells from the myotome become myoblasts is termed determination or being committed to the pathway of becoming a myoblast. Myoblasts have a single nucleus, continue to divide, and do not resemble muscle cells in that they do not contain contractile proteins. After many cycles of replication, myoblasts may stop dividing and undergo the process of differentiation into myocytes, whereby they organize themselves with each other and fuse to form multinucleated primary muscle fibers or multinucleated secondary muscle fibers, or they may remain as undifferentiated myoblasts, termed satellite cells. Primary muscle fibers appear first and provide a framework along which smaller, secondary muscle fibers will form. The number of primary muscle fibers appears to be determined by genetics, whereas the number of secondary muscle fibers formed may be influenced by nutrition and other environmental factors. Myocytes fuse with each other, resulting in a tube-like multinucleated cell with contractile proteins forming around the nuclei and along the axis of the cell. This structure is termed a myotube (Fig. 2.3) and becomes a primary muscle fiber. As more contractile proteins are synthesized and the contractile proteins become more organized into contractile units, termed myofibrils, the nuclei are forced to the outer edge of the cells and remain on the periphery of the elongated, tapered muscle cells (Fig. 2.4). Muscle cells continue to increase in number through the embryonic phase and through much of the fetal stage of development.

The early embryonic period is a critical time for development, and environmental factors such as severe heat stress of the dam can result in embryonic mortality. For example, sows and ewes exposed to temperatures of 95°F during days 10–16 of pregnancy frequently suffer loss of developing embryos. Hence, producers make a concerted effort to provide relief from heat stress to the breeding herd during this critical time. Likewise, maternal recognition of pregnancy occurs due to changes in

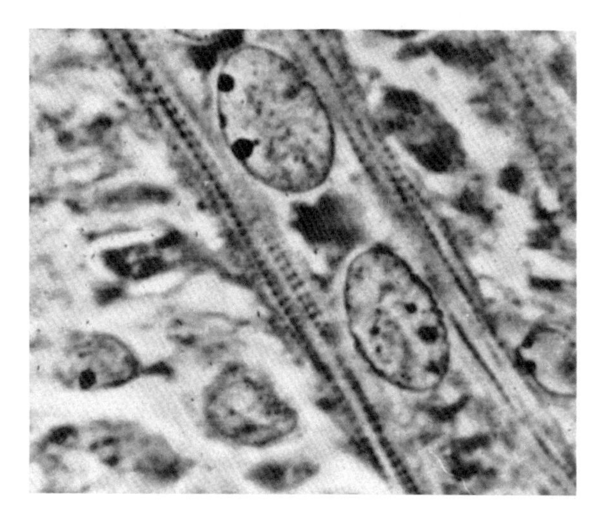

FIG. 2.3

A photomicrograph of the myotube stage of muscle cell development showing contractile proteins at the edges and nuclei in the center of the myotube.
Courtesy of H.J. Swatland, University of Guelph.

FIG. 2.4

A photomicrograph of a tapered end of a more mature muscle cell showing the striated pattern of the organized contractile proteins and multiple nuclei on the periphery of the cell. Courtesy of H.J. Swatland, University of Guelph.

hormone concentrations and ratios and results in a prioritization of nutrient use by the dam to support the developing embryo(s).

Nutrient transfer to the developing embryo is by simple diffusion from secretions of the oviduct during the ovum stage and the uterus in the early embryonic stage of development. Although the developing embryo consists more of membranes than tissue in its early stage after hatching from the blastocyst, it is obvious that diffusion cannot support the transfer of nutrients as the embryo develops. Hence, in mammals, fetal membranes develop to provide protection to the embryo as well as a mechanism whereby blood vessels from the developing embryo exist in close apposition to blood vessels in the uterus, allowing exchange of nutrients. The type of placental attachment varies among species of farm animals, but each developing embryo develops its own set of fetal membranes and placenta, regardless of the number of embryos developing in the uterus. Therefore the space available for each developing embryo can influence the amount of nutrients that can be transferred from the dam. In fact, piglets born in large litters frequently have lesser birth weights than those born in smaller litters. Likewise, nutrient restriction due to problems associated with proper development of the placenta can result in restricted growth of piglets during the fetal stage of development and birth of what are commonly called runt pigs. Regardless of efforts to provide extra nutrients to runt pigs during postnatal growth, these animals continue to grow more slowly than littermates of normal birth weight and fail to achieve a mature size equal to that of littermates (Table 2.1). Typically, runt pigs have

Table 2.1 Carcass composition of normal barrows and littermate runt pigs at 240 pounds

Item	Normal barrows	Runts
Birth weight (lb)	3.5	1.8
Age at slaughter (days)	169	188
Fat depth (in.)	1.83	1.90
Loin eye area (in.^{2})	4.73	3.74
Carcass % muscle	48.2	45.0
Carcass % fat	37.8	41.0
Carcass % bone	13.1	13.2

birth weights approximately two-thirds less than their littermates, smaller vital organs, fewer skeletal muscle cells, more carcass fat, and an inferior feed-to-gain ratio.

Efforts to increase muscle mass through improved nutrition to the dam have yielded mixed results, in that optimum nutrition of the dam results in optimum expression of the genetic potential for growth of the fetus. Providing excess nutrients to the dam late in gestation results in storage of excess body fat in the dam, which can be particularly troublesome for sows because excess body fat is frequently deposited in the birth canal, making parturition more difficult. Conversely, restriction of nutrients to the dam during gestation often results in reduced birth weight, but the dam will sacrifice nutrient stores, including muscle mass, to help ensure survival of the fetus. However, providing an improved plane of nutrition during days 20–50 of gestation, or just prior to the time of muscle fiber hyperplasia, has resulted in improved postnatal growth of pigs.

CONNECTIVE TISSUE

For our discussion of animal growth and development, it is important to discuss connective tissue of the body given that both bone and adipose tissue are actually specialized forms of connective tissue. Although connective tissue is usually thought of as being tendons, ligaments, and cartilage, connective tissue is actually the extracellular matrix of the body and provides structure and support for the body as well as a means of attachment of bones and muscles to each other. As the name implies, the extracellular matrix is outside of cells, but the proteins and carbohydrate-containing proteins known as glycoproteins are produced by adjacent cells.

Tendons provide attachment points between muscles and bones and are composed of collagen, a tough, strong complex of many protein fibers. There are many different types of collagen in the body, and discussion of their individual properties is left for other books. Tendons must be strong to transfer the mechanical energy created by muscles. Ligaments attach bones to other bones and also contain collagen as well as another type of connective tissue fiber, elastin. Elastin, as its name implies, is a somewhat elastic complex of protein fibers and allows the connections between bones to stretch slightly. A third type of connective tissue fiber is reticulin, a connective tissue protein composed of collagen but is a thin network of highly branched fibers joining adjacent types of connective tissues.

BONE

The process of osteogenesis or bone formation takes place during both prenatal and postnatal growth. Most bones have regions of hard, dense bone or softer, spongy bone. Hard, dense bone is termed cortical bone and comprises about 80% of the bone mass. Spongy, trabecular bone makes up the remaining 20% of the overall bone mass, but because spongy bone has a much more porous structure, it has approximately 10 times the total surface area of cortical bone. Hence, spongy bone is much more active metabolically.

During prenatal development, both trabecular or spongy bone and cortical bone are formed by aggregation of mesenchyme cells of the sclerotome and their subsequent differentiation into chondrocytes that produce collagen and bone matrix in a connective tissue network. Osteoblasts are bone-forming cells that are produced by chondrocytes and are responsible for deposition of minerals (primarily calcium, phosphorus, and magnesium) into the matrix. New collagen and bone matrix is formed adjacent to the previous layers of collagen in a region supplied with blood vessels and nerves. Osteoblasts trapped in the bone matrix become osteocytes, which are relatively inactive cells.

Prenatal development of the long bones of the appendicular skeleton is characterized by the aggregation of mesenchymal cells and differentiation of these cells to chondrocytes. The rudimentary structure of the bone is formed as hyaline cartilage, and an outer layer of connective tissue forms the perichondrium that surrounds the cartilage-like tissue. A central region of chondrocytes is established, and collagen in the center of the hyaline model of the bone is converted to spongy bone in a primary zone of ossification. Another zone of ossification or mineralization is established adjacent to the inner surface of the perichondrium or outer connective tissue surrounding the bone, thus forming a collar of more dense bone around the primary center of ossification. Osteoclasts begin removing spongy bone in the center of the bone, creating a bone marrow cavity, as the primary zone of ossification proceeds from the center of the bone toward each end of the bone. Additional cartilage is synthesized at the ends of the developing bone, and a secondary zone of ossification is established within the ends of the bone. Significant mineralization of the bones has occurred by the time of birth, with a high rate of cartilage synthesis continuing to occur at the ends of the bones to support continued bone growth. Prenatal development of bone requires significant deposition of minerals, and supplementation of the diet of the dam with adequate calcium, phosphorous, and vitamins is an important management priority.

ADIPOSE TISSUE

As mentioned earlier, adipose tissue is actually a specialized form of connective tissue. Adipose tissue provides shape but not structure for an animal. That is, adipose tissue does not provide structural stability or strength to an animal, but it does provide a storage area of energy for the newborn animal as well as later in postnatal life. In addition, adipose tissue provides cushioning and protection in that it is found between and within muscles and around organs. A gross examination of adipose tissue reveals a whitish color and an extensive amount of connective tissue present. Microscopic examination reveals adipose tissue is comprised of connective tissue linking many spherical cells surrounded by a thin layer of cytoplasm. In fact, these adipose cells are filled with a single mass of lipid and may increase in size by as much as 10 times, but this high degree of growth and lipid accumulation normally occurs only during postnatal growth.

Most adipose tissue described previously is termed white adipose tissue, based on its relatively white appearance. However, brown adipose tissue also exists and is usually found in immature animals rather than in mature farm animals. Brown adipose tissue is darker in color than white adipose tissue due to a higher number of mitochondria per cell. Brown adipose tissue is also composed of a number of smaller lipid droplets in the cytoplasm, and the nucleus is located more centrally than in white adipose tissue cells. The high number of mitochondria and the richer supply of blood vessels allow brown adipose tissue to oxidize stored lipids as a means of generating heat, a process of vital importance to newborn animals.

Adipose tissue cells originate from mesenchyme cells and undergo hyperplasia as preadipocytes located in a loose matrix of reticular connective tissue. Prior to differentiation from preadipocytes to adipocytes, the cells lose the ability to divide. The differentiated adipocytes then begin to accumulate the biochemical systems necessary to support lipid metabolism. Adipocytes are distinguished by the presence of small lipid droplets within the cytoplasm and the cytoplasm is forced to the periphery of the cell as lipid accumulates. The process of differentiation is under genetic control, and genes determine the characteristics of adipose tissue cells in various parts of the body. For instance, lipid stored in the perirenal region of the body contains fatty acids that are more saturated than lipid stored in the subcutaneous regions.

FETAL PERIOD

The fetal period is characterized by further development of tissues with a dramatic increase in size of the developing fetus and generally represents the final 75% of the gestation period. It is during this period that organs become functional and the animal acquires the physiological systems and controls to allow survival after birth. Bones, muscles, and connective tissues mature adequately to allow the animal to support itself and to be mobile. Muscle fiber numbers increase by the process of hyperplasia and muscle fibers group into muscle bundles. During the last one-third of the prenatal period, muscle fibers increase in size or hypertrophy, and the developing muscles fibers grow in length by the addition of complete contractile units at the ends of existing myofibrils. The number of muscle fibers becomes fixed at or near the time of birth in most species. For instance, the number of muscle fibers increases little after 90 days of gestation in the pig, marking an end to the process of muscle hyperplasia.

Some cattle display the genetic condition described as double muscling, which is a misnomer because there is no increase in the number of muscles in the carcass. There is, however, a distinct increase in the number of muscle fibers per muscle by the time of birth. Hence, the extreme muscling of such calves is readily apparent at birth.

Muscle hypertrophy is also noted in some swine, particularly those of the Pietrain breed (Fig. 2.5) as well as in some sheep. The extreme muscling in swine is present at birth, but in sheep, the extreme muscle growth begins a few weeks after birth. Again, each instance is due to extreme muscle hypertrophy, not hyperplasia, and there is no increase in the actual number of muscles in the carcass.

FIG. 2.5

Extremely heavily muscled Pietrain pigs.

SUMMARY

The time of prenatal development has been divided into three stages: ovum, embryonic, and fetal stages. The ovum stage lasts from conception until hatching of the blastocyst from the zona pellucida. The embryonic stage involves the key process of differentiation from the three germ layers of the inner cell mass of the blastocyst (endoderm, mesoderm, and ectoderm) to formation of the tissues of the organs, muscles, skeleton, and skin. The fetal stage is the time of further growth of the fetus to parturition. The endoderm becomes the lungs, liver, pancreas, and esophagus and the lining of the stomach, intestines, and colon. The mesoderm becomes the muscles, bones, and other connective tissues; heart and circulatory system; and kidneys. Myoblasts fuse together to form multinucleated myotubes and subsequent primary muscle fibers or fuse to form secondary muscle fibers adjacent to primary muscle fibers. As contractile proteins are synthesized, nuclei are forced to the periphery of the cell. The number of muscle cells is complete by the time of birth, and further increase in muscle mass is by hypertrophy. Bones are formed from connective tissue chondrocytes derived from sclerotome cells of somites. Adipose tissue also originates from connective tissue and is essentially a collection of lipid-containing cells suspended in a connective tissue network. The ectoderm becomes the brain and spinal cord, skin, and hair.

QUESTIONS FOR STUDY AND DISCUSSION TOPICS

1. Name the three stages of prenatal development.
2. Name the three germ layers that develop in the inner cell mass.

3. Name examples of different tissues that are derived from cells of the sclerotome, myotome, and dermatome.
4. What is an undifferentiated muscle cell?
5. Bone and adipose tissue are both examples of what kind of tissue?
6. What is the name of the process describing bone formation?
7. Define hypertrophy and hyperplasia. If exercise increases muscle mass, which of these processes is responsible and why?

CHAPTER

Bone growth and development with relationships to live animal and carcass evaluation

3

INTRODUCTION

The skeleton provides structural support for the body and is the organizational foundation upon and around which the muscles and organs are ultimately organized. Hence, bones assist in overall movement or support by providing a mechanical means to transfer force from muscle contraction. Bones also provide protection to sensitive organs such as the brain and the organs in the thoracic cavity. Bone marrow is a site of red blood cell production, and bone is also a reservoir for storage of minerals such as calcium and phosphorus. Therefore bone is much more than an inactive structure to which muscles are attached.

Skeletons of our domestic animals are each unique in shape and form. For instance, the skeleton of a pig is readily discernible from that of a calf, sheep, or horse, even if the skeletons of each are adjusted for differences in scale or size. It is therefore important for us to discuss the types and structure of bone comprising the skeleton, how bones grow and undergo remodeling during growth, what happens when bones mature, and how bones undergo repair in times of injury. An understanding of the principles involved in bone growth and metabolism will help us better understand how animals grow and the various changes that take place as the animal reaches maturity.

Bone is composed of an organic matrix that provides flexibility and strength, like the network of steel rods encased in the concrete foundation and structure of buildings and bridges. Approximately 30% of bone consists of this matrix, and the matrix is composed of collagen fibers and a component called ground substance, a mixture of proteoglycans and glycoproteins, and bone cells (Table 3.1). The remaining 70% of bone is composed of minerals, primarily inorganic salts of hydroxyapatites and calcium phosphate. These components are deposited in the connective tissue matrix to provide rigidity and strength. The mineral component of bone by itself would be brittle and subject to breakage. Likewise, the connective tissue matrix by itself would be flexible and unable to withstand forces exerted by muscles. However, the mineral-impregnated matrix is very strong and has the ability to remodel itself in response to changing loads as an animal grows.

The Science of Animal Growth and Meat Technology. https://doi.org/10.1016/B978-0-12-815277-5.00003-2

Table 3.1 Approximate composition of bone

Fresh bone	%
Water	45
Ash	25
Protein	20
Fat	10
Bone ash	**%**
Calcium	36
Phosphorus	17
Magnesium	<1

TYPES OF BONE

There are two primary types of bone in skeletons of domestic animals: compact or cortical bone and spongy or cancellous bone.

Compact (cortical) bone is so named because of its dense, hard nature and because the dense bone is also found as a cortical ring surrounding a bone marrow cavity of the long bones. An example would be a cross section of the femur of a pig, sheep, or calf. Cortical bone accounts for approximately 80% of the bone in the skeleton. Spongy bone is named because of its sponge-like appearance and is also described as cancellous or trabecular bone because this bone is deposited in a fibrous network. Trabecular bone is found extensively in the axial skeleton (vertebrae) as well as in the epiphyseal and marrow regions of long bones and accounts for approximately 20% of skeletal bone mass. Trabecular bone provides the major site for bone turnover because the mesh-like surface area of cancellous bone is many times greater than that of dense cortical bone.

STRUCTURE OF BONE

As shown in Fig. 3.1, the major parts of a long bone are defined as the end or epiphysis, the shaft or diaphysis, the metaphysis near the juncture of the epiphysis and diaphysis, and an area described as the epiphyseal line or plate. The epiphysis contains both compact and spongy bone, whereas the diaphysis is composed of mostly dense cortical bone with an interior bone marrow cavity and spongy bone in continuum to the growth region between the epiphysis and diaphysis. There are two important connective tissue layers in bone. The periosteum is a layer of very strong, dense, connective tissue on the outer surface of bone, and the endosteum is a layer of connective tissue lining the medullary cavity. The articular cartilage provides a smooth surface to facilitate joint movement and weight or force transfer between bones.

The cellular organization of compact bone is depicted in Fig. 3.2. Osteocytes or mature bone cells are found in concentric rings or lacunae around small blood

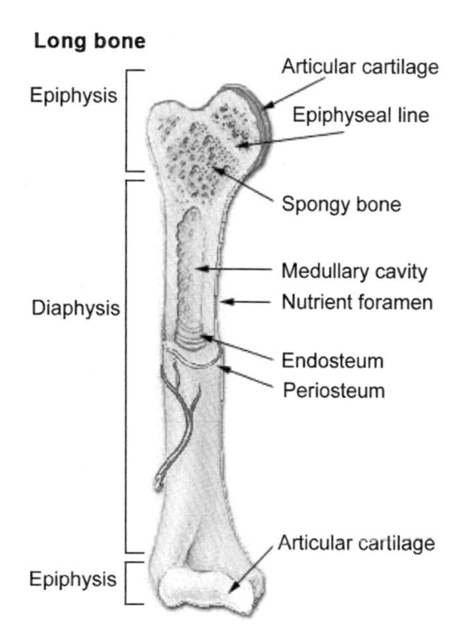

FIG. 3.1

Anatomy of a long bone.

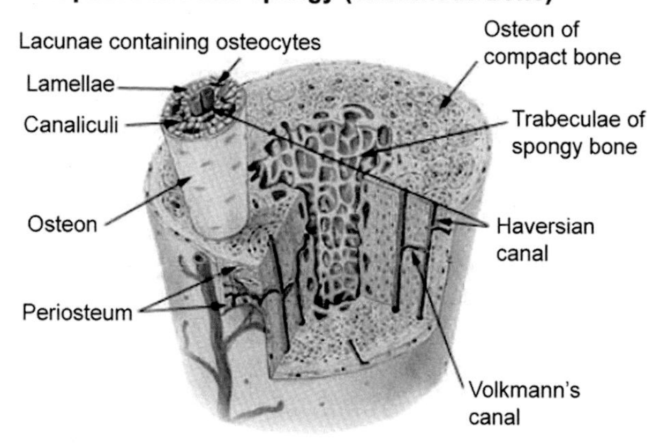

FIG. 3.2

Cellular structure of compact and spongy bone.

vessels in the Haversian canals with layers of osteon or calcified minerals between the rings of osteocytes. Osteoblasts are bone-forming cells or precursors to osteocytes and are found at the junction with spongy bone; they secrete the bone matrix or osteon. Osteoclasts are cells responsible for bone resorption or remodeling and are also found on the inner and outer surfaces of bone where bone must be reshaped.

Proper bone growth and function through maturity depends on there being a balance of activity between osteoblasts and osteoclasts.

BONE GROWTH

Bone growth takes place either by intramembranous or endochondral ossification. Intramembranous ossification is the direct deposition of bone on thin layers of connective tissue and is characteristic of the bones on the top of the skull. These intramembranous bones are formed by the evolution of mesenchyme cells to form osteoprogenitor cells which become osteoblasts. The osteoblasts secrete osteoid directly onto the connective tissue layers formed from cells of the neural crest. The osteoblasts either die or they become embedded in the bone matrix and survive as osteocytes. The resulting bone appears as distinct layers of bone.

In contrast, endochondral ossification involves the formation of bone from hyaline cartilage and is the primary method of bone formation in the body (Fig. 3.3). That is, mesenchyme cells condense and create a hyaline cartilage model of the bone to be formed. The cartilage model is infiltrated with small blood vessels and nerves, and osteoblasts begin to replace the cartilage in the central axis with spongy bone. Osteoclasts ultimately degrade spongy bone in the central area to form the bone marrow cavity, and a primary ossification center is formed. Osteoblasts and osteoclasts also form in the central region of the cartilage model under the periosteum. The osteoblasts replace cartilage with bone and a collar of ossification is formed in the center of the diaphysis. The process of ossification proceeds from the center toward the ends of the bones, and a secondary center of ossification is established in the epiphysis to allow the ends of bones to be strong enough to be able to transfer forces from one bone to another. The junction of the epiphysis and diaphysis is a zone of mitotic production of chondrocytes that ultimately produce additional cartilage. This new cartilage and its conversion to bone is the basis for longitudinal growth of bones.

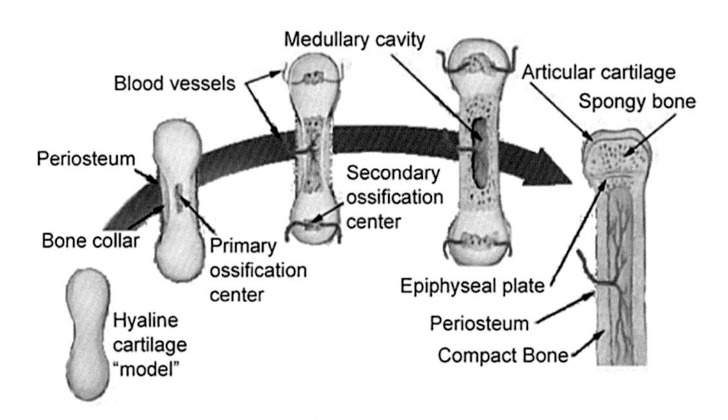

FIG. 3.3

Formation of bones by endochondral ossification.

Cortical bone of the long bones is formed by osteoblasts under the periosteum. Osteoblasts surround small blood vessels forming canals through which the blood vessels supply nutrients to the cells. The osteoblasts are organized in rings around the blood vessels and produce osteon to infiltrate the cartilage matrix. The osteoblasts ultimately die or become osteocytes, relatively inactive cells, and new rings of osteoblasts form around the blood vessels. These concentric rings of new osteocytes around the small blood vessels form the Haversian system and lacuna and are characteristic of cortical bone. The continued addition of new layers of cortical bone by this process is termed appositional growth and is responsible for the increase in width of the long bones. Osteoclasts continue to dissolve away cortical bone on the inner surface of the endosteum during growth, resulting in an increase in size of the marrow cavity and a decrease in the overall weight of long bones.

Growth of other bones of the body, such as those of the axial skeleton, is also by endochondral ossification. These bones are composed primarily of spongy bone, and growth proceeds by osteoblasts converting cartilage near the surface to bone. Osteoclasts are heavily involved in the restructuring of these bones to maintain the many channels and irregular surfaces necessary for proper function. Growth of the dorsal spine found on the thoracic vertebrae is also by endochondral ossification. The dorsal spine also increases in size by formation of bone from cartilage under the periosteum on the horizontal surfaces of the bone and by endochondral ossification of the cartilage region at the tip of the dorsal spine.

By observing the degree of ossification of the dorsal tips of the thoracic, lumbar, and sacral vertebrae, it is possible to estimate the maturity or relative age of beef carcasses. This information is one component used in determining the USDA grade of a beef carcass. As the animal matures, more of the cartilage tip becomes ossified, with complete ossification first appearing in the sacral vertebrae. Carcasses from young cattle (<24 months) will typically have white cartilage separating the sacral vertebrae and a small line of cartilage will be visible between the dorsal tip of the sacral vertebrae and the body of the sacral vertebrae. Conversely, a carcass from a mature cow (>5 years of age) will display solid bone in the sacral region because all the cartilage has been converted to bone and the sacral vertebra have become fused together (Fig. 3.4). Interestingly, the pattern of ossification is from posterior to anterior, so additional information may be gleaned by observing the degree of ossification of the thoracic vertebrae (Fig. 3.5). The thoracic and lumbar vertebrae do not become fused together with age because this would make movement difficult for the animal. However, the cartilage at the dorsal tips of the vertebrae becomes ossified in a posterior to anterior trend.

As animals mature, the mitotic activity in the epiphyseal plates slows, and existing cartilage is converted to bone, with the osteoblasts either experiencing death or becoming trapped in the osteoid matrix as osteocytes. As a result, the epiphysis becomes fused to the diaphysis and only a thin line may exist to discern where the epiphyseal plate existed. This information is used to classify lamb carcasses as either lamb (eligible to be graded in the USDA system as lamb) or mutton (not eligible for grading). Carcasses from lambs less than approximately 12 months of age will still

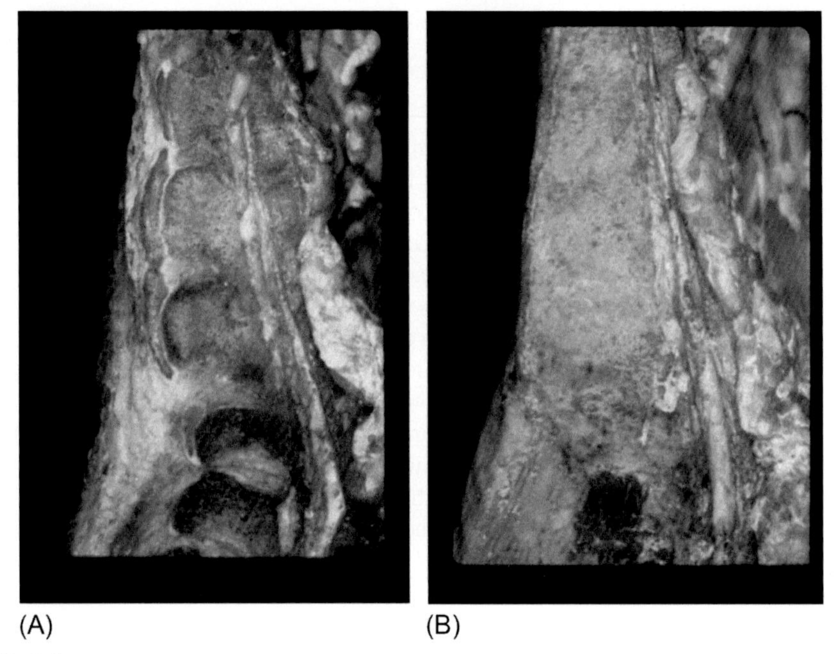

FIG. 3.4

Example of cartilage present in the sacral vertebra of a young beef carcass (A) and the fused sacral vertebra of a mature beef carcass (B).

Courtesy of the Animal Science Department, Iowa State University.

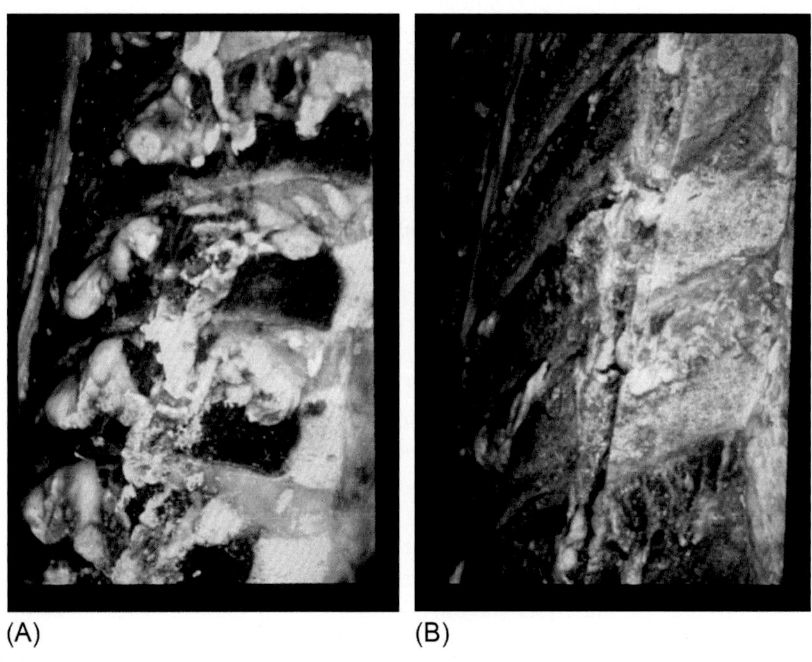

FIG. 3.5

Example of cartilage present in the dorsal tip of thoracic vertebra of a young beef carcass (A) and the ossified dorsal tip of a thoracic vertebra of a mature beef carcass (B).

Courtesy of the Animal Science Department, Iowa State University.

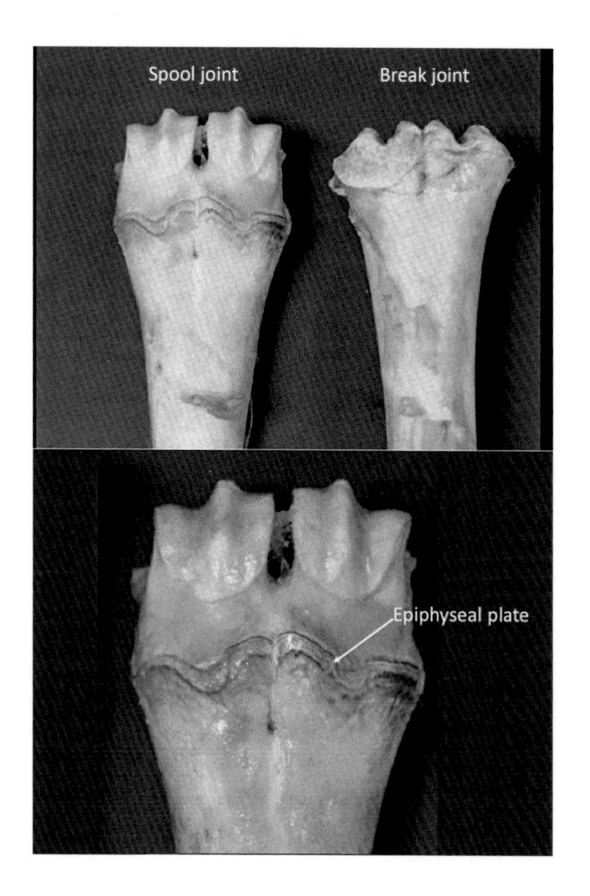

FIG. 3.6

Example of ovine metacarpals showing the spool joint and break joint of a lamb *(top panel)* and the epiphyseal plate *(lower panel).*

Courtesy of the Lonergan Lab, Animal Science Department, Iowa State University.

display the epiphyseal plate at the distal end of the metacarpal, and the foot can be separated from the metacarpal by forcibly breaking the bone at this point, hence, the term break joint, the key determining factor for carcasses to be classified as lamb (Fig. 3.6). If the joint failed to break cleanly or if it did not break at all, a spool joint would be visible and the carcass would be termed mutton. The presence of the articular cartilage on the surface of the spool joint is to be expected because this hyaline cartilage is necessary to provide a smooth bearing surface for joint movement.

BONE REMODELING AND REPAIR

During growth, shapes of bones change as forces upon them change. Just as a tree adapts to the stress of wind or load, bones also adapt to load stresses (increased weight load due to growth or less weight load due to inactivity) by remodeling the shape of the bone. For instance, the femur of a young animal will be relatively straight but may become more curved at the proximal end with subtle indentations or grooves as the weight of the animal increases and increased strain is placed on the end of the bone. Remodeling results from bone resorption by osteoclasts and corresponding synthesis of bone by osteoblasts. Osteoclasts are in contact with bone by means of their ruffled border, and the osteoclasts degrade both the organic and inorganic matrix. Osteoclasts decrease the localized pH at the ruffled border by pumping carbonic anhydrase across the border, resulting in mobilization of the inorganic matrix. The organic matrix is degraded by action of collagenase and lysosomal and nonlysosomal enzymes.

When a bone is broken, remodeling also takes place. First, damaged bone is removed by osteoclasts, and osteoblasts are recruited to synthesize new bone. The rate at which the healing process occurs is related somewhat to the relative age of the individual, in that bones of young people and young animals repair much more rapidly than those of mature individuals. Specifically, fewer osteoblasts are available or are active in mature compared with young individuals. Because bone is a much more dense tissue and is less vascular than muscle, more time is necessary for bone to heal than muscle.

Another example of bone remodeling is the situation where bone mobilization or resorption exceeds the rate of bone synthesis. This condition is termed osteoporosis in mature individuals and occurs in much of the skeleton of older humans, resulting in the loss of height as well as bone strength. Osteoporosis is not usually found in farm animals because these animals usually do not live long enough to reach such a condition. It is also likely that the diets of farm animals are better balanced for calcium, phosphorus, magnesium, and the necessary vitamins to prevent skeletal problems than are the diets of many humans.

NUTRITIONAL AND HORMONAL INFLUENCES ON BONE GROWTH AND METABOLISM

As noted in Table 3.1, bone contains predominantly calcium and phosphorus in a 2:1 ratio. Hence, most nutritional recommendations for calcium and phosphorus for growing animals approximate this ratio. In addition to these essential minerals, magnesium and vitamin D are necessary for proper bone growth. Rickets is a nutritional deficiency due to inadequate utilization of calcium, phosphorus, and vitamin D and may be found among young animals and humans while bones are growing. Osteomalacia occurs in older humans due again to inadequate utilization of calcium, phosphorus, and vitamin D and is usually cured by administration of vitamin D with adequate dietary calcium and phosphorus. Vitamin D is essential for proper calcium absorption in the intestine and is necessary for bone calcification. Hence, calcium metabolism is one of the key factors involved in proper growth. Likewise, nutritional needs continue after individuals reach their mature size because minerals are needed to support reproduction, lactation, and egg production in birds.

Calcium metabolism is a complex system coordinated by the actions of hormones produced by the parathyroid and thyroid glands. Low blood calcium concentrations result in the release of parathyroid hormone which stimulates osteoclast activity and the corresponding resorption of bone calcium. Calcium mobilized by the osteoclasts diffuses into the circulatory system and blood calcium concentrations are returned to normal. If blood calcium concentrations are too high, the thyroid gland releases calcitonin, which stimulates osteoblast activity with a corresponding deposition of calcium into bone and the blood calcium concentration returns to normal.

The thyroid gland and thyroid hormones are also important for normal bone growth because removal of the thyroid gland results in inhibition of growth. Conversely, excess production of thyroxine results in premature closure of the epiphyseal plate and cessation of long bone growth. The effects of the thyroid hormones are mediated by insulin-like growth factor-I (IGF-I), a hormone produced predominately by the liver in response to growth hormone.

Growth hormone, produced by the anterior pituitary gland, is also essential for growth, in that removal of the pituitary gland results in inhibition of growth. The effects of growth hormone are also mediated by IGF-I. The early assay for growth hormone activity was to remove the pituitary gland of rats, inject solutions thought to contain growth hormone activity into the rats, and then measure the width of the epiphyseal plate of the femur 14–17 days later. Injection of more growth hormone resulted in a corresponding increase in the width of the epiphyseal plate in the rat, and an estimate of the growth hormone activity could be made from the results.

The sex hormones estrogen and testosterone also play a role in bone growth. Estrogen functions via growth hormone and IGF-I to stimulate pubertal growth, and estrogen ultimately promotes closure of the epiphyseal plate. Chondrocytes have been shown to have both α- and β-type estrogen receptors. In females, the β form of the estrogen receptor has been shown to inhibit growth of both the appendicular and axial skeleton. Testosterone enhances the release of IGF-I and therefore also stimulates growth. Some testosterone is converted to estrogen during growth, and the α form, but not the β form, of the estrogen receptor has been shown to respond to estrogen in males. Closure of the epiphyseal plate is delayed in males compared with females due to stimulation of bone growth via androgenic effects and because the amounts of estrogen are much less in males than in females. It is also important to note that species differences exist regarding the role of specific estrogen receptors and bone metabolism, and much of what has been determined has been done comparing bone growth in rodents and humans. Castration of bulls results in a reduced growth rate, and bone growth continues at a slow rate due to the absence of estrogens. It is unusual to keep steers much beyond 2 years of age today, but this explains the very large skeletal size of oxen years ago.

SUMMARY

The two types of bone found in the body are cortical (dense) bone or spongy bone. Cortical bone is found in the shafts of long bones of the appendicular skeleton and is responsible for the strength of the long bones. Both compact and spongy bone are produced by endochondral ossification or conversion of cartilage to bone. The layered bones of the skull are produced by intramembranous ossification, a process of direct deposition of bone onto a connective tissue network. Bone is deposited by osteoblasts and mobilized by osteoclasts, whereas osteocytes are relatively inactive bone cells found isolated in the bone matrix. Longitudinal growth of long bones occurs at the epiphyseal plates, a region of chondrocyte production and cartilage synthesis at the junction of the epiphysis and diaphysis. The relative degree of ossification of cartilage of spinal vertebrae and epiphyseal plates is used as an indicator of maturity of beef and lamb carcasses, respectively. Bone growth is stimulated by growth hormone, thyroid hormones, and IGF-I, and bone growth is inhibited by estrogens and a deficiency of vitamin D. Bones undergo dramatic remodeling during normal growth and in response to added mechanical stress as well as during periods of inactivity.

QUESTIONS FOR STUDY AND DISCUSSION TOPICS

1. Name the two major parts of a long bone.
2. Name the growth region of a long bone.
3. What is ground substance?
4. Name the two primary types of bone and their respective locations in bone.
5. Name the three types of bone cells and identify their respective functions.
6. What is osteoporosis?
7. How is the process of bone ossification used to determine the relative age of beef and lamb carcasses?

CHAPTER

Muscle growth and development and relationships to meat quality and composition 4

INTRODUCTION

The increase in muscle mass during postnatal growth is due to an increase in the size of individual muscle fibers (hypertrophy) because the increase in the number of muscle fibers (hyperplasia) ceases during the late prenatal period, as discussed in Chapter 2. The number of muscle fibers present at birth and their subsequent growth are influenced by genetics and environmental conditions, such as nutrition of the dam, nutrients available to the growing animal, and the physical environment of the animal. The purpose of this chapter is to discuss how muscles hypertrophy as the animal grows from a neonate to market size and how these changes alter the shape and size of animals.

MUSCLE HYPERTROPHY

Although the number of muscle fibers or cells is already determined near the time of birth, the number of nuclei within the cell continues to increase as the muscle grows. In fact, the increase in the number of nuclei is essential for growth of the muscle fiber. The nucleus is responsible for production of RNA and other cytoplasmic organelles necessary for protein synthesis and cellular metabolism. As the cell grows in volume, more nuclei are needed to produce adequate amounts of RNA to support cellular growth, thus maintaining a relatively constant ratio of cytoplasmic volume per nucleus. Because nuclei in the multinucleated muscle cell are unable to divide, nuclei are added by the replication of satellite cells located between the basement membrane and the sarcolemma. The daughter cell produced by the division of a satellite cell fuses with the muscle cell, thereby adding an additional nucleus to the muscle cell. This process is obviously repeated many times during growth of the animal and is limited by genetic factors controlling muscle size.

Satellite cell activity is greatest while the animal is growing because there is less need for additional nuclei after the muscle cell reaches its mature size. However, satellite cells remain present to support cellular needs in cases of injury or induced hypertrophy resulting from training programs developed, for example, for humans and race horses. Satellite cells pass through specific stages of reproduction characteristic of other cells described as the cell cycle (Fig. 4.1). When there is little or no stimulation to reproduce, satellite cells remain suspended in a subsection (Gap 0, $\mathbf{G_0}$) of the Gap 1 ($\mathbf{G_1}$) state and have very little cytoplasmic material present. When

The Science of Animal Growth and Meat Technology. https://doi.org/10.1016/B978-0-12-815277-5.00004-4

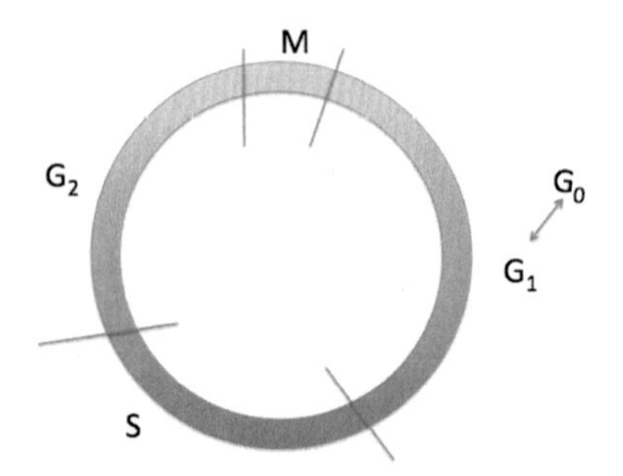

FIG. 4.1

The cell cycle shown in stages defined as G0 (Gap 0, resting state), G1 (Gap 1, initial cell growth stage of the cycle), S (synthesis of DNA), G2 (Gap 2, additional cell growth and preparation for cell division), and M (mitosis).

stimulated by specific growth factors, satellite cells become activated and begin to grow in size during the G_1 state. During the synthesis (**S**) stage, DNA synthesis occurs and additional cell growth takes place during the G_2 stage. Finally, mitosis (**M**) and cell division occurs, and one or both daughter cells are fused into muscle cells or may remain as a satellite cell.

The process of muscle growth is under genetic control, and muscle growth is a dynamic process involving both the continued synthesis and breakdown of muscle proteins. The maximum rate of muscle growth is limited by genetic factors, whereas the actual rate of muscle growth is a function of the rate of protein turnover (the rate of protein synthesis less the rate of protein breakdown). Muscle growth and size could be thought of as a volume of water in a bucket having a small hole in the bottom. If the rate at which water flows into the bucket is faster than the water leaves through the hole, the volume of water in the bucket increases, as in muscle growth. Likewise, if the rate at which water leaves the bucket is the same as the rate water enters the bucket, the volume of water stays constant, much as is the case when an animal reaches its mature size and there is no further increase in muscle size. If the rate that water leaves the bucket is faster than the rate that water enters, then the volume of water in the bucket will decrease, as happens to muscle size when animals may not receive adequate nutrition or if animals simply get old.

Animals with a greater genetic propensity for growth have a greater rate of muscle protein synthesis during growth than animals of the same species with a lesser genetic propensity for growth. An example of these genetic differences is seen by comparing growth rates of broiler chickens selected to produce muscle and growth rates of layer chickens selected to produce eggs. Modern broilers reach market weight in roughly half the time that was needed 50 years ago, and the amount of muscle produced per day of age is much greater for modern market pigs compared with their ancestors 50 years ago.

The rate of protein synthesis begins to decline as animals achieve approximately half their mature size, provided they have access to a suitable environment including a diet that supports optimum growth. At maturity, animals frequently loose muscle mass, as is often seen in older cattle, swine, and sheep (and humans). When increased environmental demands are placed on reproducing females, muscle components can be utilized to support the nutrient needs for lactation or a developing fetus during pregnancy. In these instances, the rate of protein breakdown exceeds that of protein synthesis.

Satellite cells are also an important source of nuclei when muscle has been injured and protein synthesis must be stimulated. Following muscle injury, damaged muscle tissue is removed and new muscle protein must be synthesized. Satellite cells appear soon after the injury and removal of damaged tissue to aid the regeneration process. The overall healing process requires support of the nervous system, and injured muscle often fails to heal properly if the nerve supply to the muscle has also been damaged. Similarly, when nerves supplying muscles are damaged by disease or injury, the affected muscles often atrophy as a result of lack of use. Again, atrophy is the result of the rate of protein breakdown exceeding the rate of protein synthesis in the affected muscle fibers.

The proteins present in muscle are constantly being broken down, synthesized, and assembled in the live animal during growth and repair of muscle. Three major protease systems are responsible for most of the breakdown of proteins in muscle fibers. These proteases are found within the cell and include the lysosomal system, the calpain system, and the proteasome. The lysosome is an organelle within the cell containing several proteases, but these proteases function best in very acidic environments, and hence, do not have a major role in proteolysis of living muscle. The calpain system is comprised primarily of two calcium-dependent proteolytic enzymes, mu-calpain and m-calpain, and calpastatin, a modulator of calpain activity. The calpains degrade specific proteins in muscle fibers and liberate protein fragments from the larger protein structure. The role of the calpain system in postmortem muscle metabolism and meat tenderness will be discussed in detail in Chapter 10. The proteasome is the major structure in the cell responsible for degrading previously liberated proteins and protein fragments to short polypeptides.

Myofibrils also degrade postmortem because some of the mechanisms responsible for degradation of muscle tissue continue to function after blood circulation ceases. In fact, these processes are important in determining the ultimate tenderness of meat. Mu-calpain helps break down proteins in the Z-disk of postmortem muscle, thereby disrupting the highly organized muscle structure, which can ultimately influence meat quality and reduce the effort needed to shear or chew cooked meat. The activity of mu-calpain in postmortem muscle varies among muscles within a carcass as well as from carcass to carcass. As mentioned previously, lysosomes function better in the lower pH environment of postmortem muscle and they also contribute to postmortem breakdown of muscle fibers.

MUSCLE FIBER TYPES

Muscle fibers have been classified into three types based on their physiological and metabolic characteristics, and the distribution of these fiber types varies both within and among muscles. Slow twitch, oxidative fibers or Type I fibers are characterized by a slow contraction speed and the ability to utilize oxygen for the production of energy for contraction (ATP). Hence, Type I fibers have a relatively high supply of capillaries to the muscle fiber. Fast twitch fibers (short contraction time) are composed of two types: those that rely primarily on anaerobic (glycolytic) metabolism to produce energy (Type II β) and those that have the capability to utilize both oxidative and glycolytic metabolism to produce energy (Type II α). Muscles typically contain all three types of fibers, but the relative distribution of fiber types varies among specific muscles. Muscles with a greater proportion of slow twitch fibers are typically more red in color than muscles containing a higher proportion of fast twitch fibers due to a higher content of myoglobin (oxygen-binding protein in muscle) in slow twitch fibers. Lighter colored muscles are typically composed primarily of fast twitch, glycolytic muscle fibers, with a lesser proportion of fast twitch, oxidative fibers and slow twitch, oxidative fibers. These fast twitch, glycolytic fibers rely primarily on the anaerobic breakdown of glycogen to produce energy. Perhaps the best examples of the distribution of fiber types can be found in birds. Migratory birds have flight (breast) muscles that must contract and relax for hours at a time and therefore rely on oxidative pathways for energy production. These muscles are typically dark red in color compared with the relatively white breast muscle of the domestic chicken that flies for only a very short distance, if it flies at all. Likewise, the leg and thigh muscles of the domestic chicken are red and are used for sustained periods of standing or for slow locomotion. Although the uses and color differences among muscles are not as dramatic in pigs and cattle compared with chickens, many of the leg muscles of pigs and cattle are typically redder than their loin muscles. Many different characteristics have been used by researchers to describe muscle fiber types and their function, and a summary of these traits is presented in Table 4.1.

During growth, muscles that contain predominately slow twitch, oxidative fibers grow and mature with those same characteristics. Muscles with a greater proportion

Table 4.1 Selected characteristics of types of muscle fibers

Type I	Type II β	Type II α
Slow contraction speed	Fast contraction speed	Fast contraction speed
Aerobic metabolism	Anaerobic metabolism	Anaerobic and aerobic metabolism
Highest content of myoglobin	Least amount of myoglobin	Intermediate amount of myoglobin
Highest content of fatty acids	Least amount of fatty acids	Intermediate amount of fatty acids
Smallest fiber diameter	Largest fiber diameter	Intermediate fiber diameter
Low concentration of glycogen	High amount of glycogen	Intermediate amount of glycogen
Red color	White color	Intermediate (pink) color
Rich blood supply	Least blood supply	Intermediate blood supply

of fast twitch fibers have a greater ability to adapt to the environmental needs of the animal. That is, the metabolic function of those fast twitch fibers with the ability to generate energy through both aerobic and anaerobic metabolic pathways can be induced to become more aerobic or anaerobic depending on the conditions experienced by the animal. An animal exposed to greater than normal aerobic activity can become more effective in that capacity. Likewise, an animal exposed to strenuous activity for short periods of time will become more effective in that capacity. However, no amount of training can overcome the limitations imposed by genetic control of muscle development. Simply put, some animals within a species are faster than others and are more powerful than others, and these differences are largely genetic. For example, athletes who excel as sprinters are seldom effective as marathon runners, although sprinters and marathon runners can enhance their respective abilities through various training programs.

Mature muscle fibers also vary in diameter over their length and taper at the ends of each fiber. Muscle fibers with the largest central diameter are those having fast twitch, glycolytic properties, and the fibers with the smallest central diameter are the slow twitch, aerobic fibers. Hence, the fibers with intermediate metabolic properties also have intermediate cross sectional areas. Animals selected to have larger muscles have muscles with an increased number of muscle fibers and also have muscles with more fast twitch, glycolytic properties. This has resulted in a general deterioration in meat quality due to the greater number of muscle fibers with glycolytic properties and an increased rate and amount of postmortem glycolysis in some muscles of high consumer value. The result in the pork industry is the development of a condition known as pale, soft, and exudative (**PSE**) pork, and the pork loin and interior muscles of the ham are the primary muscles affected. PSE pork is light (almost white) in color, soft in texture, and exudes cytoplasmic fluids on the cut surface of affected muscles. This meat has a tendency to be less juicy and more tough after cooking than meat of normal grayish pink color, firm texture, and only slightly moist to the touch on the cut surface of a muscle [red, firm, and normal (**RFN**)]. If pigs predisposed to become PSE after slaughter are exposed to mild, long-term stressors before slaughter (extreme hot weather or relocation to an unfamiliar, stressful environment for a few days before slaughter), the stores of muscle glycogen may be reduced by the time of slaughter. As a result, there is little substrate for postmortem glycolysis and the meat has a dark red (reddish-brown) color, a very firm texture, and is dry (almost sticky) to the touch of a fresh cut surface of the muscle (**DFD**). Pork that is DFD has limited appeal in the retail case and is typically routed to the production of processed meats. An intermediate condition of red, soft, and exudative (**RSE**) pork has also been described but is of lesser importance to that of PSE, a major problem for the pork industry since the 1960s. Examples of PSE, normal, and DFD pork are shown in Fig. 4.2.

A related condition of dark, firm, and dry beef, known as dark-cutting beef is caused by long-term stress imposed on cattle and the subsequent metabolism of muscle glycogen stores before slaughter. Dark-cutting cattle are most often found among cattle slaughtered following shows at fairs or cattle from feedlots when there has been long-term exposure to extremely hot, stressful conditions or exposure to

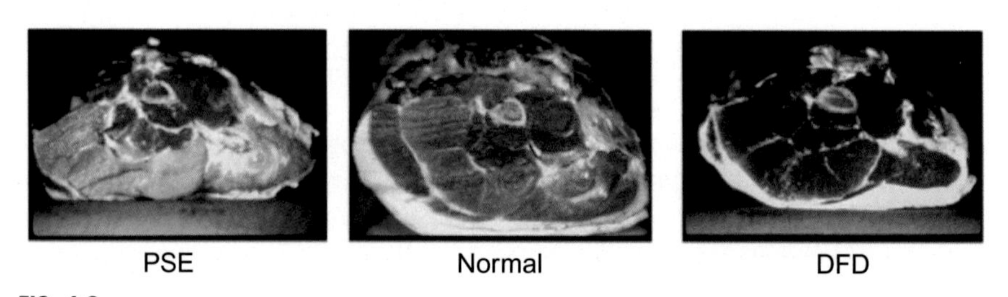

FIG. 4.2

Examples of PSE, RFN, and DFD pork muscle in the cut surfaces of hams.

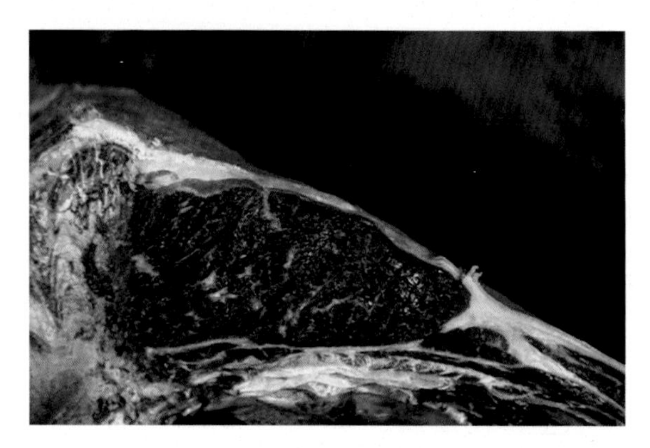

FIG. 4.3

An example of dark-cutting beef in a carcass from a market animal.

rapid, major weather changes such as during a change of season. An example of dark-cutting beef is shown in Fig. 4.3.

CHANGES IN MUSCLE SIZE DURING GROWTH

Postnatal growth of muscle is characterized by changes in both length and girth of muscles and corresponding muscle fibers. The increase in muscle fiber diameter occurs by incorporation of additional contractile proteins on the periphery of myofibrils (longitudinal subunits of muscle fibers) and longitudinal splitting of myofibrils to create additional myofibrils. This process continues until the mature size of the muscle is reached. Exercise can also induce muscle hypertrophy by the production of additional muscle mass. Hence, stretching muscles under load is an effective method to increase muscle strength and diameter. Sustained aerobic training can induce greater working capacity of muscle but there is little change in muscle fiber diameter. However, prolonged anaerobic training, such as done by weight lifters, results in an increase in muscle mass due to an increase in diameter of muscle fibers, specifically the intermediate type (fast twitch with both glycolytic and aerobic metabolism). In growing animals, the increased load of greater body mass also serves as a stimulus for muscle growth.

Muscles grow in length by the addition of whole sarcomeres (repeating contractile units along the length of muscle cells) at the ends of myofibrils and occurs while allowing muscles to contract and relax during normal use. The increase in muscle length is necessary to maintain the relationship between muscles and the skeleton or tissues to which muscles attach. Because the ends of muscle fibers are a key region of protein synthesis, a larger proportion of nuclei are found at the ends of growing muscle fibers compared with the central portion of fibers. However, the distribution of nuclei along the remaining surface of muscle fibers is relatively uniform, as would be expected to maintain normal cell function, including the synthesis of muscle proteins involved in radial growth of muscle fibers.

Faster growing breeds of animals typically have more fibers in the same muscle and hence more muscle mass than animals in slower growing breeds. The sex of an animal also influences the rate of growth and resulting muscle mass. For instance, boars have more muscle mass and grow faster than littermate gilts or barrows, and gilts grow faster with more muscle mass than littermate barrows. Similarly, bulls sired by the same male grow faster with more muscle mass than heifers and steers sired by the same male, and steers typically grow faster with more muscle mass than half-sibling heifers. The observation that castrate male calves grow more rapidly with less body fat than intact females whereas intact female swine grow more rapidly with less body fat than castrate males has yet to be explained. The greater muscle mass of intact males compared with genetically related castrated males is most likely due to the hypertrophic effects of the androgenic hormone testosterone.

Domestication and genetic selection of swine has resulted in an increase in the number of muscle fibers and an increase in rate of growth compared with wild boars. Likewise, domestic turkeys are dramatically more heavily muscled than wild turkeys. Selection for muscle mass in cattle has resulted in cases of extreme muscularity in some mature animals. Cattle that are extremely heavily muscled have been termed "double muscled," and examples of this condition may be found among some breeds, including the Belgian Blue, Charolais, Maine-Anjou, and Piedmontese. Double muscling is due to a mutation of the myostatin gene and the effect is expressed during the prenatal period. Cattle with this condition display extreme muscling at the time of birth and the condition is evident throughout life (Fig. 4.4). These cattle have a greater number of muscle fibers and their muscle fibers are larger in diameter than those from normal cattle. Double-muscled cattle that have been selected for meat production grow faster and have heavier muscles than cattle selected for milk production. The muscle-to-bone ratio of cattle increases dramatically from 4.1 for dairy cattle to 5.1 for beef cattle and to 6.1 for double-muscled cattle.

An extreme muscling condition among sheep is known as "callipyge" which translates to "beautiful buttocks" from Greek. This condition is due to a genetic mutation traced to one male, and expression of this gene takes place during growth after approximately 3 weeks of age (Fig. 4.5). The heavy muscling seen in the hind saddle of the callipyge lambs appeared to be desirable from a live-animal standpoint, and lambs with this muscular appearance were selected and promoted. However, selection for the callipyge genotype resulted in a product with reduced consumer acceptance. Muscle from callipyge lambs is less tender than muscle from normal lambs and is

FIG. 4.4

Example of a double-muscled Belgian Blue bull. Note the definition of musculature in the rear quarter.

Photograph courtesy of the American Belgian Blue Breeders Inc.

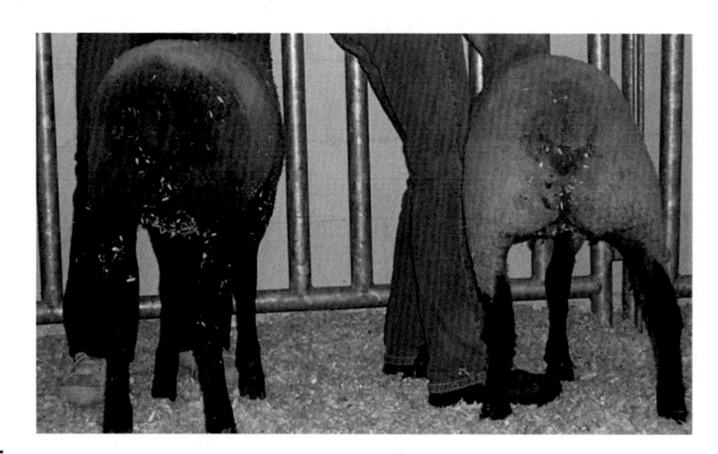

FIG. 4.5

Examples of normal *(left)* and callipyge *(right)* lambs.

Courtesy of Steven Lonergan, Iowa State University.

likely due to less postmortem proteolysis in the callipyge muscle. Therefore genetic tests have now been developed to help eliminate the condition from the lamb industry.

SUMMARY

Muscle growth is the result of protein synthesis, and the rate of muscle growth slows as the animal reaches approximately half its mature size. Muscle fibers or muscle cells have been classified based on their physiological and metabolic characteristics. Muscles contain a mixture of three types of muscle fibers: Type I (red), Type II β (white), and Type II α (intermediate). Red muscle fibers are used for sustained work, such as long-term locomotion, and are termed aerobic because they derive their energy from oxidative metabolism. White muscle fibers are larger in diameter than red fibers, are used for short-term, rapid bursts of activity such as quick movement, and are termed anaerobic because they derive their energy from the metabolism of glycogen. Intermediate muscle fibers utilize both oxygen and glycogen to generate ATP for muscle contraction. Selection for animals with larger muscles results in animals having a larger proportion of white and intermediate-type muscle fibers. Pigs with a larger proportion of white muscle fibers are prone to experience a more rapid rate of postmortem metabolism resulting in poor quality meat that is pale, soft, and exudative (PSE). Likewise, meat from cattle selected for extreme muscle development (double muscling) have muscles that are lighter in color than normal cattle.

QUESTIONS FOR STUDY AND DISCUSSION TOPICS

1. Why are satellite cells important for muscle growth?
2. When does muscle growth begin to decline?
3. Proteins are broken down in muscle by what three protease systems?
4. List the three types of muscle fibers and how each generates energy.
5. Provide examples of how muscle use determines the predominant type of muscle fiber present.
6. Define the terms PSE and DFD.
7. Why are some cattle described as double muscled?

CHAPTER

Fat and fat cells in domestic animals

5

INTRODUCTION

Fat cells (adipocytes) have been viewed for many years as passive cells and their major function was thought to be the storage of excess energy. Recent research, however, has established that fat cells have other functions in addition to the storage of energy. Fat cells also function as active endocrine cells that produce several proteins and peptides that communicate directly and indirectly with the brain and peripheral tissues. This allows the fat cell to function as an interface between energy intake and the regulation of body fat deposition in different parts of the body. That is why fat cells are strategically positioned in the animal's body to make up the intramuscular fat (marbling), intermuscular fat (seam fat), subcutaneous fat (outside fat of the carcass), and internal fat (inside the body cavity).

In normal animals, fat cells produce hormones as well as molecules that carry very important information regarding the regulation of the size of the fat tissue reserves and the regulation of specific peripheral and central pathways needed to control proper energy balances in domestic animals. In this chapter, information will be presented to reflect the specific functions of fat cells on obesity development and the selection for the reduction of body fat depots in domestic animals to meet the demands for reduced fat in retail and wholesale cuts of pork, beef, and lamb.

To understand the controls for fat deposition, it is also important to understand the regulation of growth and development of fat tissue. Therefore this chapter will also include information on fat tissue at different growth stages.

GROWTH AND DEVELOPMENT OF FAT CELLS AND THEIR RELATIONSHIP TO BODY COMPOSITION

When an animal is born, its body contains only a small percentage of chemical fat (lipid). Lipids are soluble in organic solvents and not soluble in water. The chemical composition of a newborn pig is presented in Table 5.1 as an example for newborn domestic animals.

ADIPOSE TISSUE CELLULARITY

When an animal is born, more fat cells that contain lipids are located within the body tissues, such as muscle, and very few lipid-containing fat cells are in the outer portion

The Science of Animal Growth and Meat Technology. https://doi.org/10.1016/B978-0-12-815277-5.00005-6

Table 5.1 Chemical composition of a newborn pig

Component	Percent chemical composition
Lipid	2.7
Protein	14.7
Water	77.6
Ash	5.7

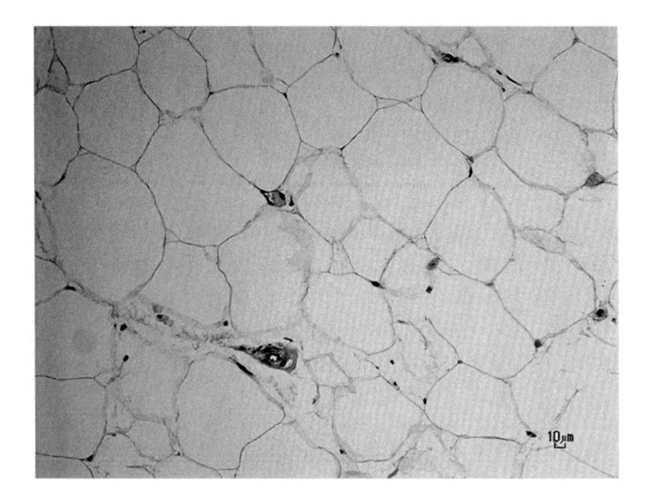

FIG. 5.1

An example of fat cells (adipocytes).

Courtesy of Dr. Elizabeth Whitley, Department of Veterinary Pathology, Iowa State University, Ames.

of the body, such as subcutaneous fat tissue. Therefore the 2.7% chemical fat of the newborn pig is located within tissue and is not often observed outside of the newborn carcass. When an animal grows after birth, fat deposition occurs by enlargement of individual fat cells and by the addition of new fat cells in adipose tissue. Examples of fat cells, adipocytes, are shown in Fig. 5.1. Fat cells from brisket and subcutaneous fat depots are shown in Fig. 5.2.

In general, adipocytes from younger steers exhibited a hexagonal-like shape (Fig. 5.2B) and changed to a spherical-like shape as cells were filled with more storage lipid (Fig. 5.2D). The number of adipocytes per gram of adipose tissue decreased with age in all fat depots (Fig. 5.3). The rate of decrease in the number of cells per gram of adipose tissue during growth was greater before 15 months of age. From 15 to 19 months of age, the number of cells per weight of tissue decreased more slowly or remained constant. Conversely, the average diameter of the adipocytes in the six different fat depots increased as steers grew to 17 months of age (Table 5.2).

The enlargement of individual fat cells accounts for most of the fat deposition as an animal grows. The fat cells can increase as much as six times in diameter

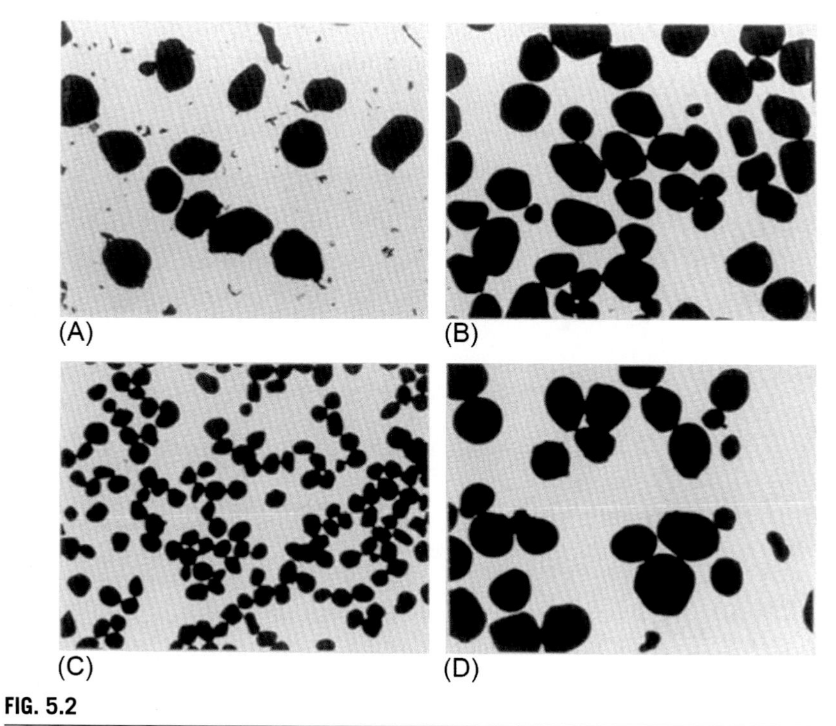

FIG. 5.2

Examples of adipocytes from beef cattle. Adipocytes in sections (A) (×75) and (C) (×30) are from subcutaneous adipose tissue of 19-month-old steers. Adipocytes in (B) (×75) and (D) (×75) are from brisket adipose tissue from 11- and 19-month-old steers, respectively.

Courtesy of the Animal Science Department, Iowa State University.

when calorie intake exceeds expenditures during postnatal development. When this size is reached, more fat cells will develop for additional deposition of lipid in the fat tissues of thc body. An example of the separable fat deposition as well as muscle and bone percentage in different growth stages of the pig is presented in Table 5.3.

Forty billion fat cells are common in tissues of the body in a normal adult. When high levels of fat occur, there can be 100 billion fat cells in an individual who is obese. During growth and development of animals, a chemical compound called leptin is produced by the fat cells along with chemical messengers that are also produced by the fat cells. These chemical messengers have critical physiological functions and act as inflammatory and antiinflammatory agents that have major effects on tissues throughout the body. They can also influence heart disease in people and animals. Therefore during the growth process, fat cells have a very significant effect on other tissues in the body and also regulate their own lipid accumulation within the cell.

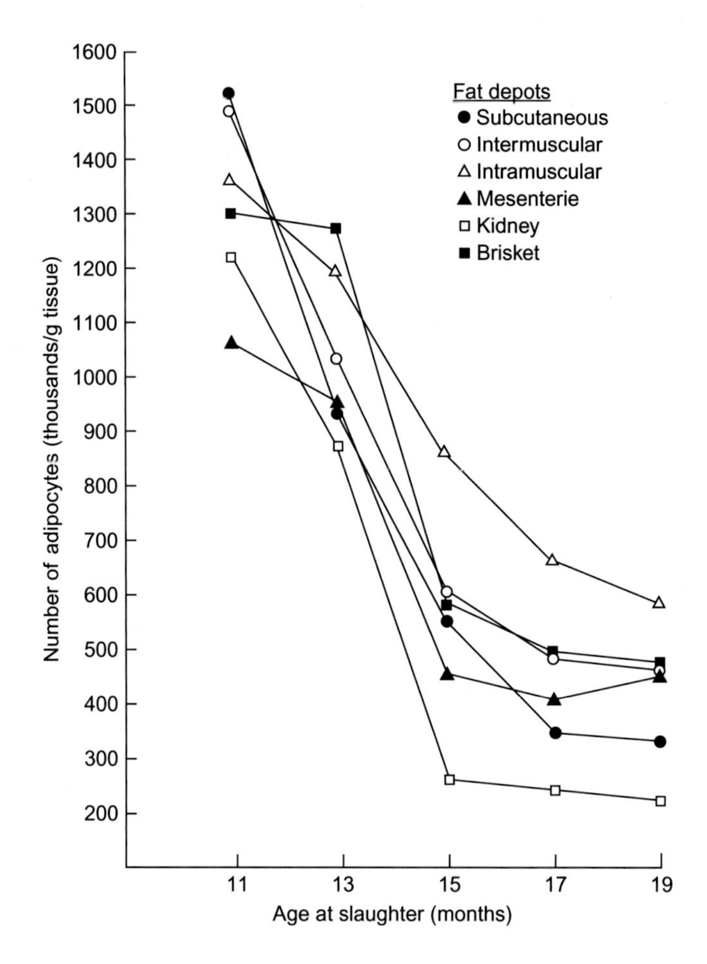

FIG. 5.3

Number of fat cells per gram of adipose tissue as a function of age in six fat depots of steers.

Courtesy of the Animal Science Department, Iowa State University.

Table 5.2 Average diameter of fat cells (adipocytes; measured in μm) in fat depots[a]

	Age at slaughter (months)			
Fat depot	**11**	**13**	**15**	**17**
Subcutaneous	95	120	122	138
Intramuscular	73	86	96	107
Mesenteric	103	117	145	149
Kidney	93	122	148	183
Brisket	74	96	117	116

[a] *Values are means of eight steers per slaughter group.*

Table 5.3 An example of the physical composition of pigs at different growth stages

Live weight (pounds)	Percent carcass		
	Fat	Muscle	Bone
44	12	67	21
110	18	65	17
198	25	62	13
264	29	60	11

ANATOMICAL LOCATION OF FAT TISSUE DURING GROWTH AND DEVELOPMENT

Development of fat tissue occurs at specific sites in the animal during the growth and development process. The sites are called fat depots and they are formed from the accumulation of lipid in the fat cells (adipocytes), resulting in an increase in the cell size. There are several fat depots in meat-producing animals and they are listed in Table 5.4 for beef cattle. The intermuscular fat is greater than the subcutaneous fat in each age group. Kidney fat and mesenteric fat had similar values as the cattle increased in weight from 11 to 19 months of age.

VISCERAL FAT DEPOSITION

The visceral fat depot is one of the first depots to be developed. It is called visceral fat because it is located close to the viscera. The major physical function of the visceral fat is protection of the internal organs. It also provides a reserve source for energy if needed by the animal.

Table 5.4 Amount of dissectible fat tissue (measured in pounds) in selected fat depots for steers at 11–19 months of age

Fat depot	Chronological age of steers at slaughter (months)				
	11	13	15	17	19
Subcutaneous fat	23.5	34.5	60.2	86.6	95.0
Intermuscular fat	42.9	65.5	107.1	139.7	155.5
Brisket fat (external)	2.0	2.2	4.6	8.1	10.5
Kidney fat	8.6	12.9	22.6	27.3	34.1
Mesenteric fat	8.6	14.5	22.4	22.4	31.6
Total dissectible fat					
Pounds per steer	97.6	151.1	249.9	315.9	373.3
Fat thickness, 12th rib (in.)	0.20	0.30	0.32	0.51	0.55

Courtesy of the Animal Science Department, Iowa State University.

PELVIC, HEART, KIDNEY, AND SUBCUTANEOUS FAT IN CATTLE

The pelvic, heart, and kidney fats are other examples of internal body fat and these fat depots are used in the Yield Grading Standards in addition to the subcutaneous fat for USDA Yield grades to predict beef carcass value. Examples of pelvic, kidney, and heart fats are shown in Fig. 5.4. Kidney fat surrounds the kidney, is present even in young animals, and increases greatly as mature weights are reached. The kidney fat can protect the kidney from bruises, so it has an important physical function. An example of subcutaneous fat in the beef carcass is shown in Fig. 5.5. Subcutaneous fat is located under the hide of cattle and it is one of the major fat depots. It is used as an indicator of total fat in the carcass, and the 12th–13th rib subcutaneous fat location is one measurement used in the USDA Yield Grade Standards.

SUBCUTANEOUS FAT DEPOSITION IN PIGS

The subcutaneous fat depot is one of the major depots to be developed during the growth process of the pig. Subcutaneous fat is located under the skin of animals and accounts for the largest amount of fat in the pork carcass. Subcutaneous fat is deposited in layers, and connective tissue separates the three layers of subcutaneous fat. Fig. 5.6 shows the three layers of fat in a fat-type pig. As an animal grows, the

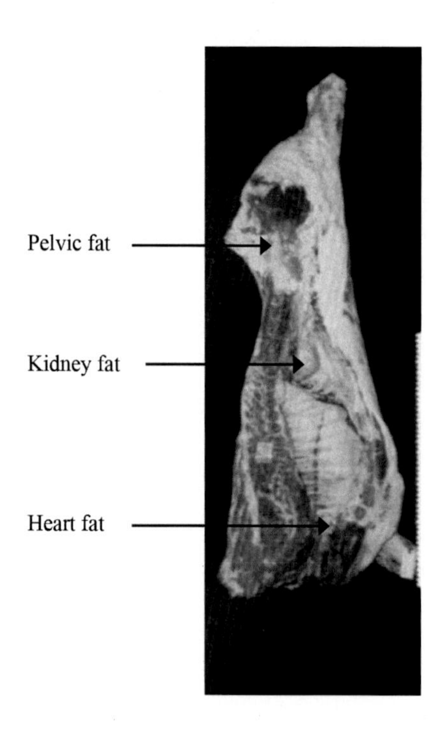

FIG. 5.4

Example of pelvic, kidney, and heart fat in the beef carcass.

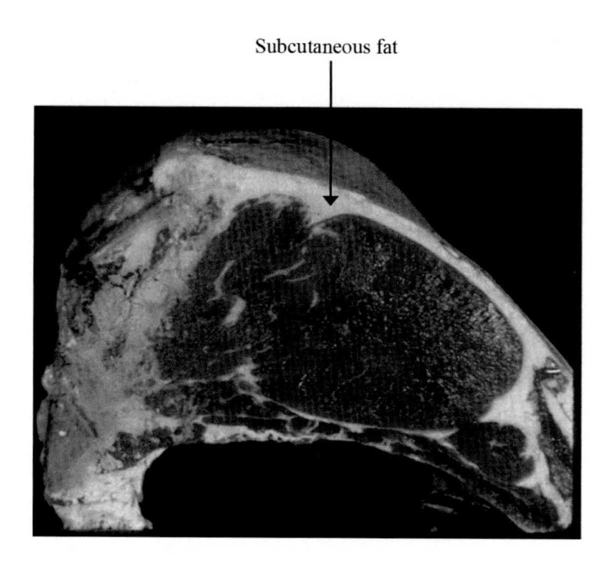

FIG. 5.5

An example of subcutaneous fat over the rib-eye muscle (longissimus dorsi) in the beef carcass (12th–13th rib section).

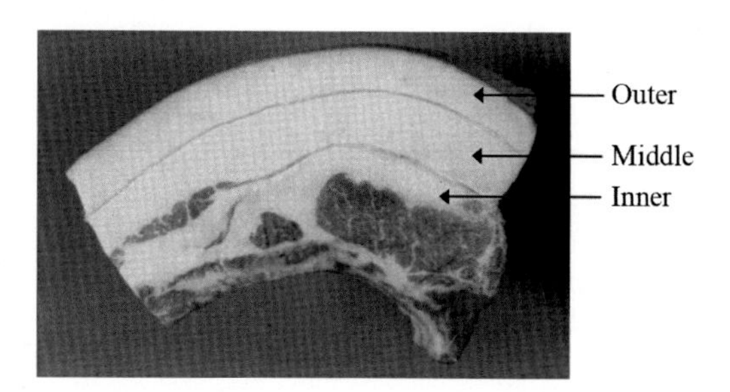

FIG. 5.6

Illustration of the three layers of subcutaneous fat over the loin-eye muscle in the pork carcass of a very-fat-type pig.

three subcutaneous fat layers are clearly differentiated. The outer layer is laid down or deposited relatively early in the growth process and increases much less than the middle backfat layer as the animal grows and develops to a mature weight.

The layer of fat directly over the longissimus muscle (loin eye muscle) develops at a later stage of the growth process when compared with the outer and middle layers. In very trim and muscular pigs, the inner layer of subcutaneous fat is very small or may not even develop (Fig. 5.7). Table 5.5 reflects the relative difference in layers of subcutaneous fat in the loin region of the pig. The inner fat layer reported in Table 5.5 combines the middle fat layer and the inner fat layer of subcutaneous fat shown in Fig. 5.7.

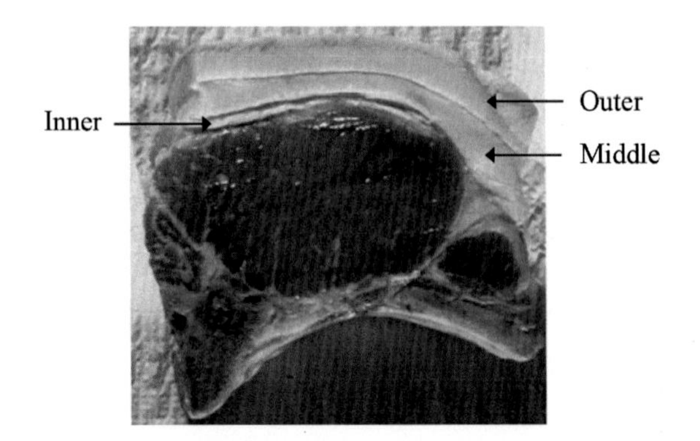

FIG. 5.7

Illustration of the three subcutaneous fat layers over the loin-eye muscle in a meat-type pig.

Table 5.5 Relative changes in layers of subcutaneous fat in the loin region of pigs

Fat location	Age (weeks)			
	Birth	4	16	24
Outer layer (in.)	0.06	0.17	0.24	0.37
Inner layer (in.)	0.06	0.27	0.44	1.00
Total (in.)	0.12	0.44	0.68	1.37

From McMeeken, C.P. Growth and Development in the Pig. J. Agric. Sci. Camb. 31, 1–49, 1941.

In the selection of meat-type pigs, swine breeders have reduced the middle layer of the subcutaneous fat considerably. The breeders have also reduced the outer layer of subcutaneous fat but not to the degree of the middle layer. This comparison is shown in Fig. 5.7.

It is apparent that the outer layer of subcutaneous fat is more essential to the pig than the middle layer for physiological functions and adjustments to the environments as it is more difficult to change the thickness of this layer by genetic selection or environmental adjustments such as ration energy levels.

SUBCUTANEOUS FAT DEPOSITION IN CATTLE

Examples of subcutaneous fat deposition patterns for Angus steers weighing 850–1000 pounds are shown in Figs. 5.8 and 5.9. It should be noted that the greatest fat deposition at 850 pounds is in the rump, hind flank, lower loin, brisket, and center of the shoulder. The fat deposited in these areas helps to protect the animal from bruises as they pass through gates or other small openings. During the growth of Angus cattle from 850 to 1000 pounds, a large proportion of the fat is deposited in the lower flank. Figs. 5.10 and 5.11 reflect the proportion of subcutaneous fat in the flank region of cattle for meat- and fatter-type cattle as they reach market weight. Only a few

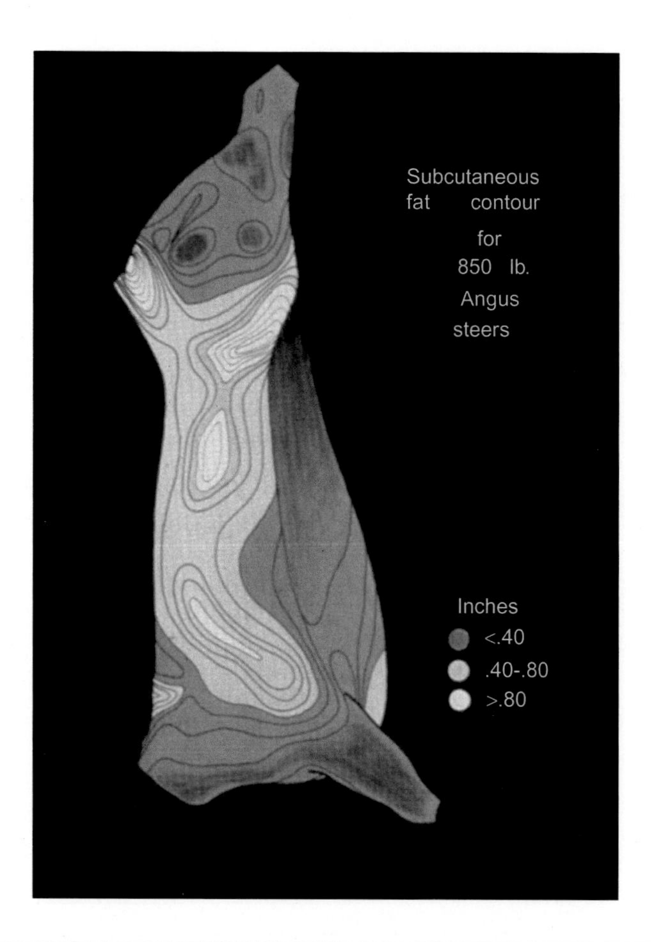

FIG. 5.8

An example of the subcutaneous fat deposition contours in a carcass from an 850-pound Angus steer.

Courtesy of the Department of Animal Sciences, University of Wisconsin-Madison.

very-fat-type cattle exist in today's market, but the emphasis to reduce subcutaneous fat in some USDA Choice cattle is still needed.

INTERMUSCULAR FAT

Intermuscular fat is located between muscles and often surrounds moving muscle surfaces. Examples for beef carcasses are shown in Figs. 5.12 and 5.13. It also fills the space between bone and points of muscle attachments. As pigs grow, intermuscular fat increases at a faster rate than subcutaneous. One of the physical functions of intermuscular fat is to act as a buffer and reduce friction from muscle movement, particularly in the legs, neck, and thorax. Also, other major functions of intermuscular fat as well as fat in other fat depots are associated with physiological effects. Most of

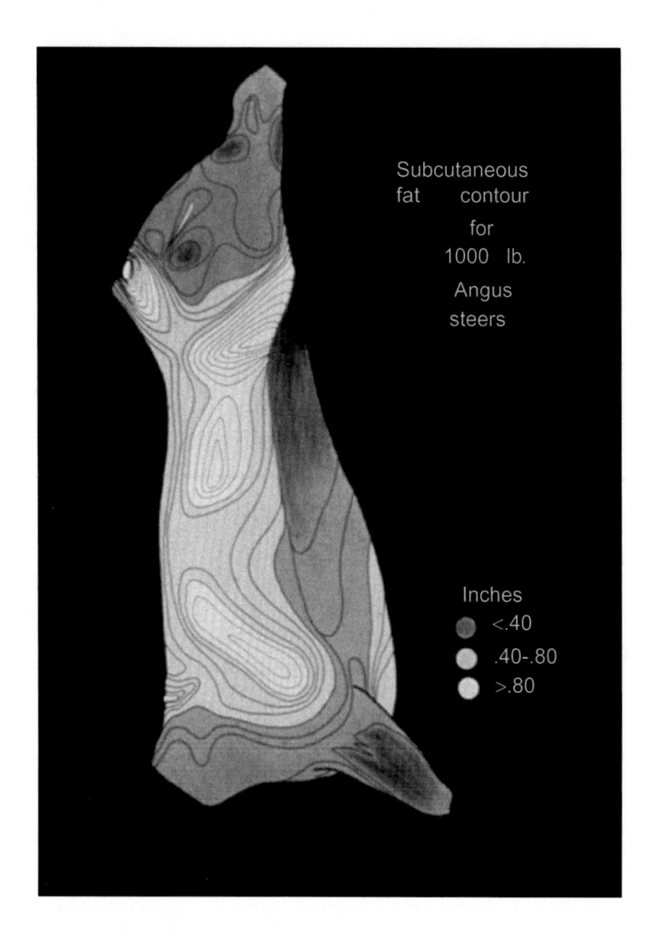

FIG. 5.9

An example of the subcutaneous fat deposition contours in a carcass from a 1000-pound Angus steer.

Courtesy of the Department of Animal Sciences, University of Wisconsin-Madison.

the scientific information in this area is from humans and experimental animals such as mice and rats. The physiological functions of intermuscular fat as well as other fat depots takes place within the fat cells in the fat depots. The fat cells function like chemical factories as they are continually absorbing or releasing substances such as resistin and leptin in response to the body's energy needs. Recent research with domestic animals, particularly the pig, is contributing to the fundamental understanding of fat cell functions and fat deposition in the growth stages of meat-producing animals. This is a new and exciting area for researchers working in the animal sciences. The pig is much like humans for physiological functions, so the pig is a good model for human medical research on fat development. Recent research results obtained using pigs as the experimental animal confirms the fat cell research conducted with humans is similar to the physiological functions of fat cells in pigs.

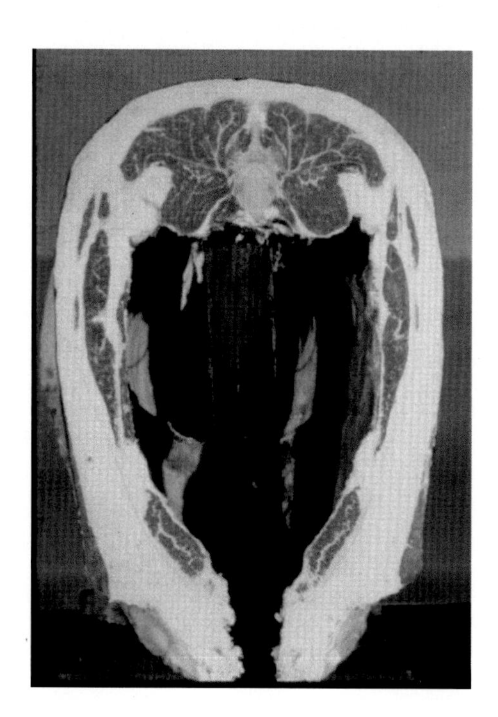

FIG. 5.10

An example of a large amount of fat in the flank region of a carcass from a very fat steer.

From E. Kline and R. Taylor. Courtesy of the Animal Science Department, Iowa State University.

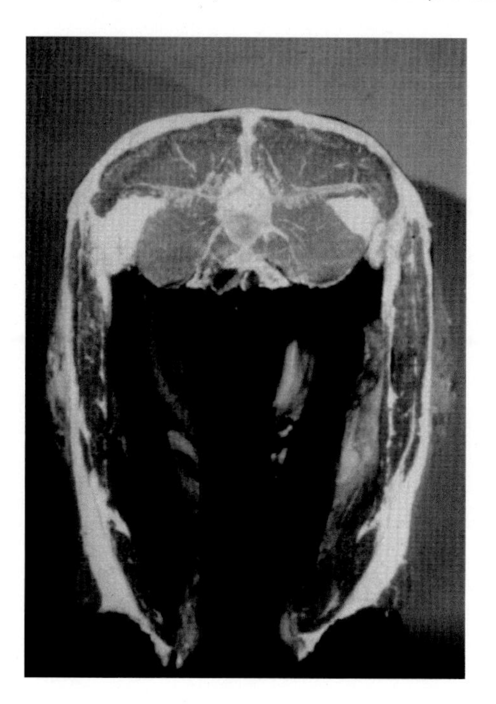

FIG. 5.11

An example of fat in the flank region of a carcass from a meat-type steer.

From E. Kline and R. Taylor. Courtesy of the Animal Science Department, Iowa State University.

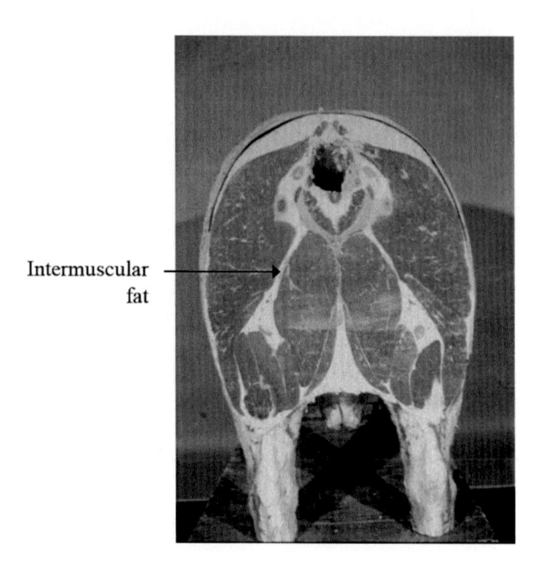

FIG. 5.12

An example of intermuscular fat in the cross section of the round portion of a beef carcass.

From E. Kline and R. Taylor. Courtesy of the Animal Science Department, Iowa State University.

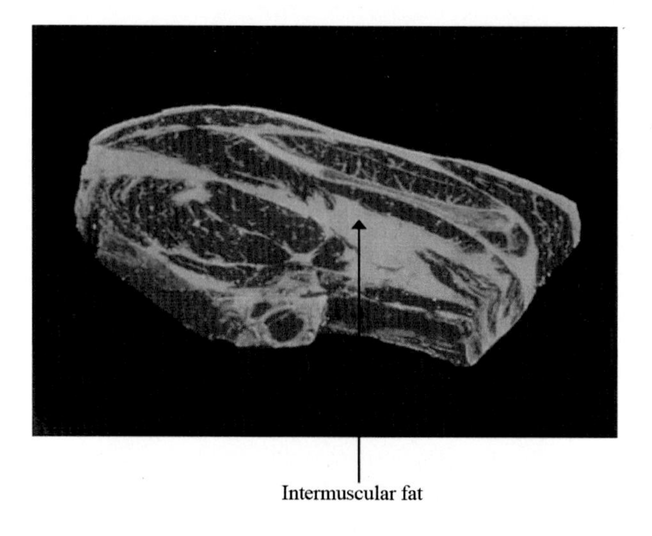

FIG. 5.13

An example of intermuscular fat in the chuck section of the beef carcass.

INTRAMUSCULAR FAT (OR MARBLING)

Intramuscular fat (Fig. 5.14) is located between and within muscle fibers (cells) and its greatest deposition is in the later stages of the growth process. Intramuscular fat is called marbling in the meat industry and marbling has a significant impact on marketing fresh meat, particularly beef and pork loin cuts. The higher grades receive higher prices. The degree of marbling is used in the USDA Beef Grading System. The USDA marbling specifications have the highest degree of marbling for the USDA Prime grade and lower amounts of marbling for the USDA Choice and Select grades (Fig. 5.15). An illustration of the amount of marbling within 6 marbling degrees of the USDA Quality Grades for beef carcasses is shown in Fig. 5.16. The Abundant Marbling and the Moderately Abundant Marbling represents the USDA Prime grade; the Moderate, Modest, and Small degree of marbling represents the USDA Choice grade; and the Slight Amount of marbling represents the USDA Select grade. The relationship between the degree of marbling and the percentage of intramuscular fat is presented in Table 5.6. Marbling is also important for export standards used for pork sold to Japan and other Asian nations. Fig. 5.17 shows marbling standards used by exporters of quality pork cuts from the United States. Japan importers of US pork will pay a premium for highly marbled cuts. The same marketing concepts for marbling apply to beef exported to Japan.

VARIATION OF MARBLING BETWEEN ANIMALS

Breeds of animals have much variation in their genetic traits for marbling. For example, the Duroc and Berkshire breeds have lines with a high degree of marbling when compared with other swine breeds. It is also important to know that not all Berkshire

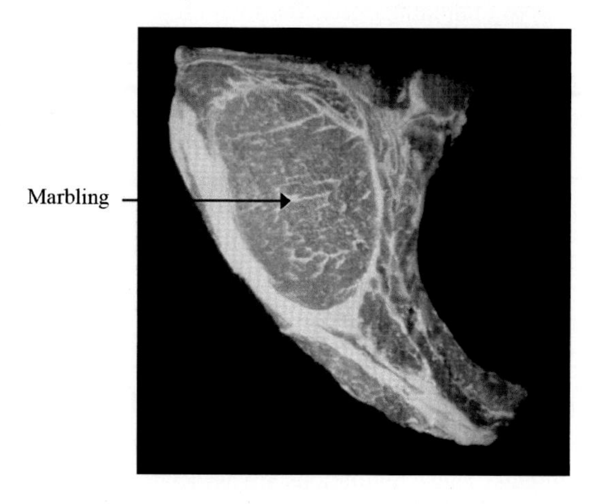

FIG. 5.14

An example of intramuscular fat in the pork muscle from the loin region of the carcass.

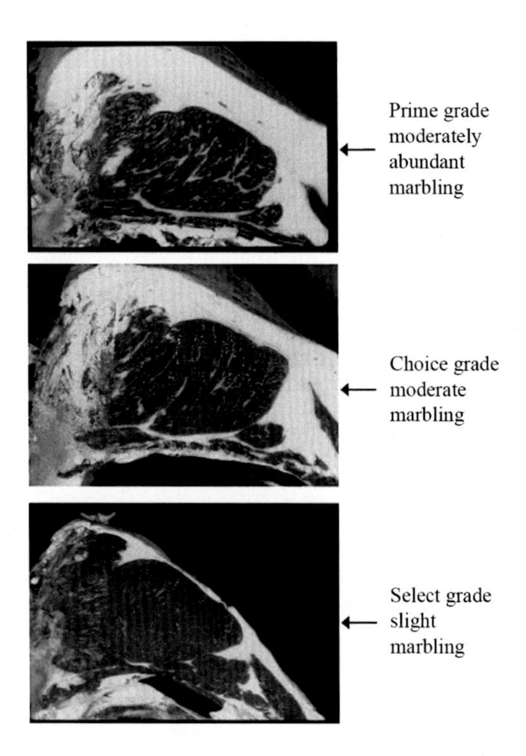

FIG. 5.15

An example of the three levels of intramuscular fat in beef rib-eye muscle: (12th–13th rib) moderately abundant (Prime), moderate (Choice), slight (Select).

and Duroc lines have a high degree of marbling. Therefore selection programs are needed within breeds to maintain genetic lines with high marbling traits.

Within purebred beef breeds, Angus have an excellent genetic base for marbling when compared with other beef breeds. When dairy breeds are fed for meat production, the Holstein breed has more genetic potential for marbling than the other dairy breeds.

The Holstein breed can deposit enough marbling to reach the Choice or even Prime grade if fed a high-energy ration for a long feeding period.

Figs. 5.18–5.20 show beef cattle that are representative of the Prime, Choice, and Select grades.

FACTORS THAT INFLUENCE MARBLING DEPOSITION

Genetic regulations of animal growth are the major controls for marbling deposition in domestic animals. These regulations are associated with the growth stages of the animal. Marbling is the last type of fat to be deposited when the growth stages are compared. When an animal reaches maturity (4th growth stage), most of the

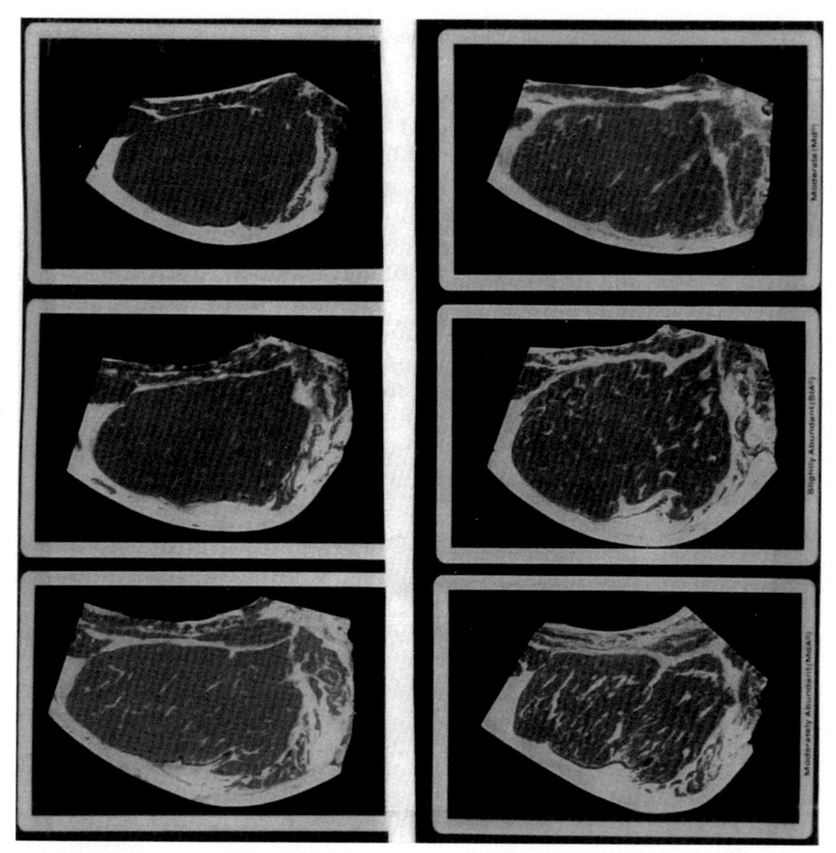

FIG. 5.16

Marbling standards used for the USDA Beef Quality grades. Left column *(top down)*: slight, small, modest. Right column *(top down)*: moderate, slightly abundant, moderately abundant.

Courtesy of the USDA.

Table 5.6 Relationship between percentage of intramuscular fat, marbling score, and carcass quality grade in beef cattle

Grade	Marbling score	Percentage intramuscular fat
Prime +	Abundant	
Prime °	Moderately abundant	12.3 and higher
Prime –	Slightly abundant	9.9–12.2
Choice +	Moderate	7.7–9.8
Choice °	Modest	5.8–7.6
Choice –	Small	4.0–5.7
Select +	Slight +	3.1–3.9
Select –	Slight –	2.3–3.0
Standard +	Traces	2.2 and lower
Standard °	Practically devoid	
Standard –	Practically devoid –	

From Gene Rouse, Courtesy of Animal Science Department, Iowa State University.

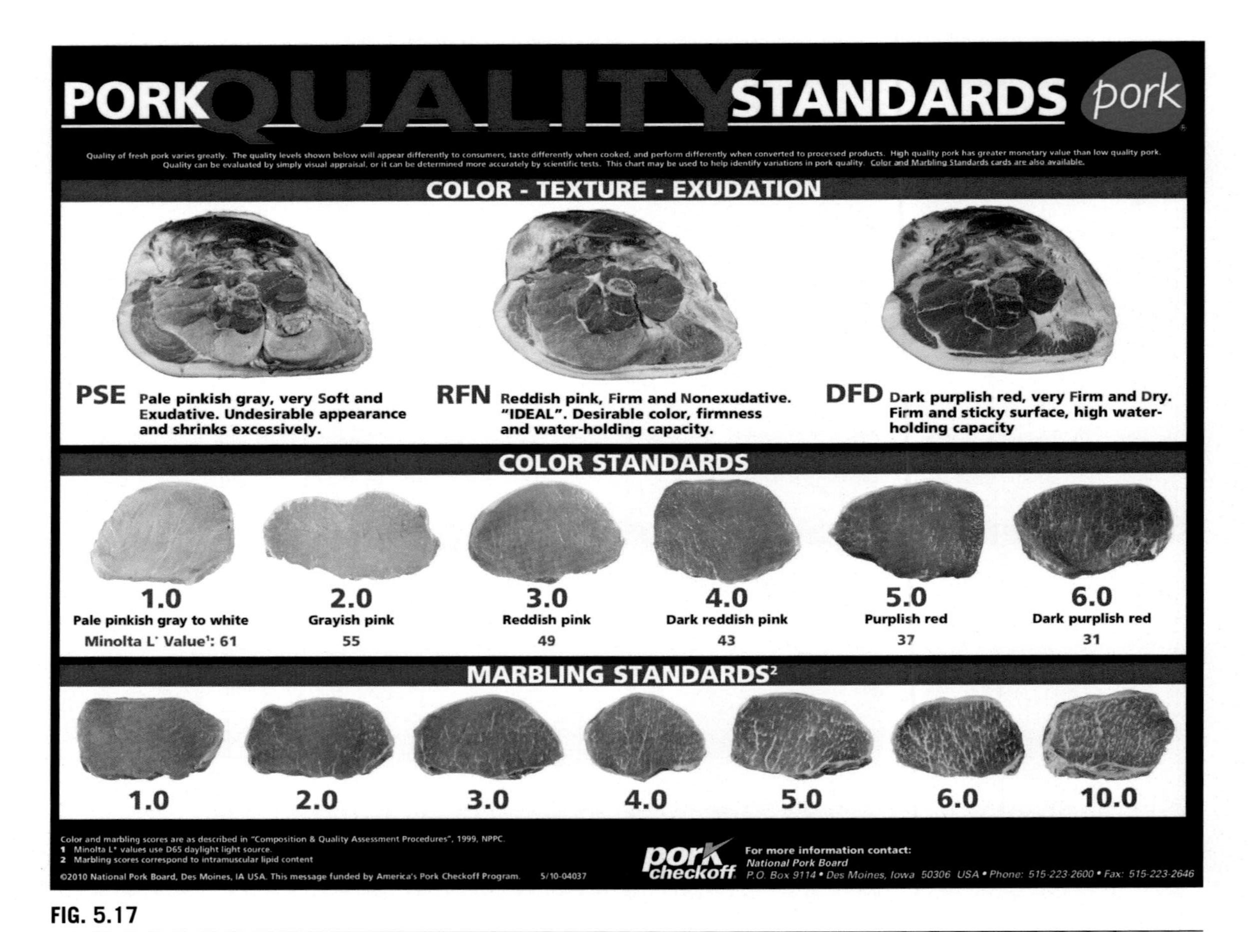

FIG. 5.17

An example of marbling standards used for the selection of pork for export.

Courtesy of the National Pork Board, Des Moines, IA.

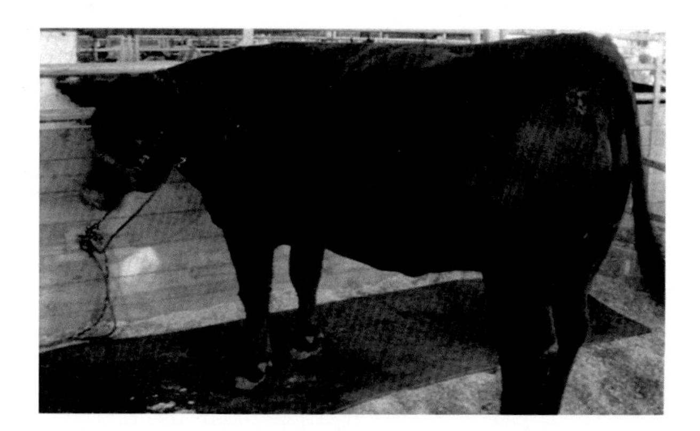

FIG. 5.18

An example of beef cattle representing the Prime grade.

FIG. 5.19

An example of beef cattle representing the Choice grade.

FIG. 5.20

An example of beef cattle representing the Select grade.

marbling will be deposited. A high proportion of the marbling starts to be deposited in the beginning of the 3rd growth stage (see Fig. 6.8, Chapter 6).

The sex of the animal interacts with animal weight to influence the rate of fat and marbling deposition in domestic animals. In pigs, boars have less fat as well as marbling than gilts and gilts have less fat and marbling than castrates (barrows).

When beef cattle are compared, bulls have less fat and marbling than steers and steers have less fat and marbling than heifers. Hormones have a major effect on fat and marbling deposition when heifers and steers are compared.

CHEMICAL COMPOSITION OF ADIPOSE TISSUE

The chemical composition of fat tissue is variable and can change with age and species. A general range is presented in Table 5.7.

The lipid components of fat tissue in domestic animals are predominantly neutral lipids and they are in the form of fatty acids located in the fat cells which are mostly found in the fat depots of animals. The chemical composition of fatty acids in the fat depots determines the firmness or hardness of carcass fat. Considerable variation exists between species for the hardness of fat. Lambs have the hardest fat, cattle rank second, and pigs have much softer fat than lambs or cattle. The higher the concentration of saturated fatty acids, the harder the fat. Therefore pigs have a much higher concentration of unsaturated fatty acids in their fat tissue when compared with lambs and cattle. When retail cuts are frozen and held in a freezer for several months, the harder the fat (higher concentration of saturated fatty acids) the greater the stability for quality traits and reduced rancidity development.

ANATOMICAL LOCATIONS AND FAT TISSUE TRAITS

The fatty acid components in fat tissue are associated with the anatomical location of fat in the carcass and the growth stage.

In the pig, the outer subcutaneous fat layer has more unsaturated fatty acids than the inner layer and the inner layer has more unsaturated fatty acids than the marbling or intramuscular fat. Comparatively, the fat around the kidney in pigs is the most saturated of the four fat types.

Table 5.7 Chemical composition of adipose tissue

Item	Percent
Lipid	76–94
Protein	1–4
Water	5–20

As the animal grows, the proportion of fatty acids change to more saturated fatty acids, resulting in a harder fat tissue in the carcass.

The degree of saturation of fatty acids, and hence the types of fatty acids in the fat tissue, influences the flavor, storage stability, and baking traits of food as well as meat processing traits for sausage production. This explains why specific fats have application in different food processing industries and why beef, pork, and lamb have different flavors after cooking.

SUMMARY

Development of fat tissue occurs at each of the four growth stages, but the majority of fat tissue is deposited in the end of the third and fourth growth stages. The fat is deposited in a highly organized order through genetic control mechanisms. Subcutaneous fat, intermuscular fat, and intramuscular fat are examples of the building blocks for the fat depots. When adipocytes (fat cells) are developed and enlarged, they are filled by the synthesis of fatty acids. This process takes place in each growth stage. The adipose tissue is a very active physiological tissue that produces substances that regulate fat deposition and fat reduction. Resistin and leptin are examples of hormonal control substances released from the fat tissue that control fat deposition. The rate of fat deposition is also influenced by genetics, the rate of animal growth, energy and nutrient intake, efficiency of growth, and animal age. The fat components of the carcass can influence meat flavor positively and negatively, meat processing, meat marketing, and to some degree, meat tenderness. Various types of fat in the carcass are very important factors for the economic base for profits in the animal industry.

QUESTIONS FOR STUDY AND DISCUSSION TOPICS

1. What are the cellular functions of fat cells (adipocytes) as pigs grow from birth to a mature weight?
2. How many total fat cells can be in the body of an obese steer?
3. List the fat depots in pigs and describe the locations of the fat depots.
4. How does marbling content in beef cattle rib muscle influence the USDA Quality grade?
5. In what growth stage is the most marbling deposited in cattle?
6. When sheep, pigs, and cattle are compared, what species has the hardest (most saturated fat), and does the hardness of the fat influence the storage stability when frozen?

CHAPTER

Growth curves and growth patterns

6

INTRODUCTION

This chapter will include fundamental concepts related to the shape of the growth curve and concepts related to physical changes in body and carcass composition during growth of animals. The emphasis will be placed on pigs, cattle, and sheep. An understanding of growth curve patterns and concepts will provide useful information for the application of the principles for animal growth that can be used for production practices, such as genetic selection, and estimating efficiency of weight gain and carcass value. Many concepts for animal growth, however, can result in very complicated patterns, and some fundamental growth control mechanisms still need more study before they can be applied to practical production practices. Information in this chapter will also explain and use applied concepts for growth curves and their application to animal production.

DEFINITIONS FOR GROWTH CURVES

Growth is often described as an increase in live weight gain per unit of time. Growth can also be defined as a progressive increase in size (length, height, girth, volume) or weight of an animal during a specific time period. Some scientists define growth in two phases. The first definition is measured in mass (weight) per unit of time, as described previously. The second definition involves changes in form and body composition resulting from differential growth of the body components. All of the definitions have a good science base and often individuals will select a definition for growth based on the application for a specific growth comparison. From the definitions presented, growth curves, therefore are usually developed by plotting factors used for weight or volume against units of time. The most common growth curve plots bodyweight against days of the animal's age.

Growth can also be associated with an accretion of nutrients in the body over the lifetime of an animal because nutrient levels control the growth rate as well as the expression of genetic regulators for the growth process. These factors can also be plotted for growth curves and patterns.

The Science of Animal Growth and Meat Technology. https://doi.org/10.1016/B978-0-12-815277-5.00006-8

CONCEPTS RELATED TO GROWTH CURVES

The general shape of a growth curve from birth to a mature weight is shown in Fig. 6.1. When weight is plotted against time units, a sigmoidal curve is obtained when an animal grows from birth to a mature weight. This shape reflects an acceleration of growth at puberty and a significant slowing of growth as maturity starts to occur.

A comparison for growth traits of cattle, swine, and sheep is shown in Fig. 6.2. Note that the general shape of the growth curve is the same for cattle, sheep, and pigs, but when they start to reach a more mature weight, the species differences in weight are evident. Also, the chronological age at maturity is different between species.

Variation in growth rate is observed within species and between breeds so the relative growth rate becomes an important economic factor for efficiency of gain and profit potential. A comparison of two dairy breeds is an example, and their growth patterns are plotted in Fig. 6.3. Holstein and Jersey breed comparisons at each growth phase reflect major differences in weight. The feed required for growth and maintenance is proportional to the differences shown in the growth curve for weight at a specific age. These differences are directly related to production costs and the economics of dairy production. The shape of the curve related to growth rate is highly regulated by genetic potential for growth as well as nutrient intake. Therefore growth curve data can be used to estimate genetic merit of the parents. The same concepts can be applied to growth curves for beef cattle, sheep, and pigs.

Faster gaining animals will have a more extended growth curve and they will mature at a heavier weight. Slower gaining animals will have a growth curve that starts to level at a lighter weight, reflecting early maturing traits. Three Herefords

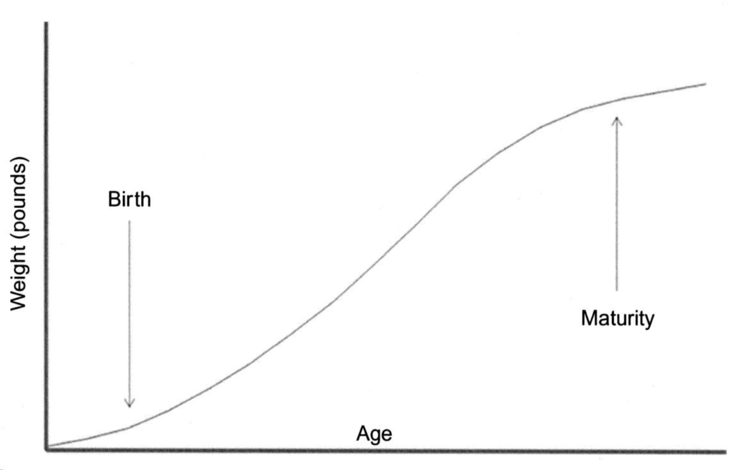

FIG. 6.1

An example of the general shape of a growth curve from birth to a mature weight in domestic animals.

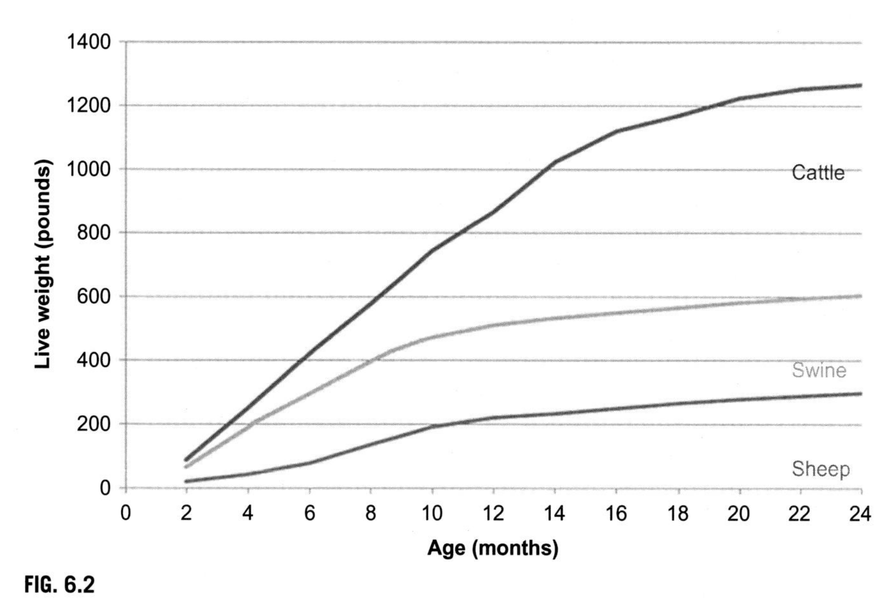

FIG. 6.2

An example of growth curves for cattle, swine, and sheep.

Courtesy, Animal Science Department, Iowa State University.

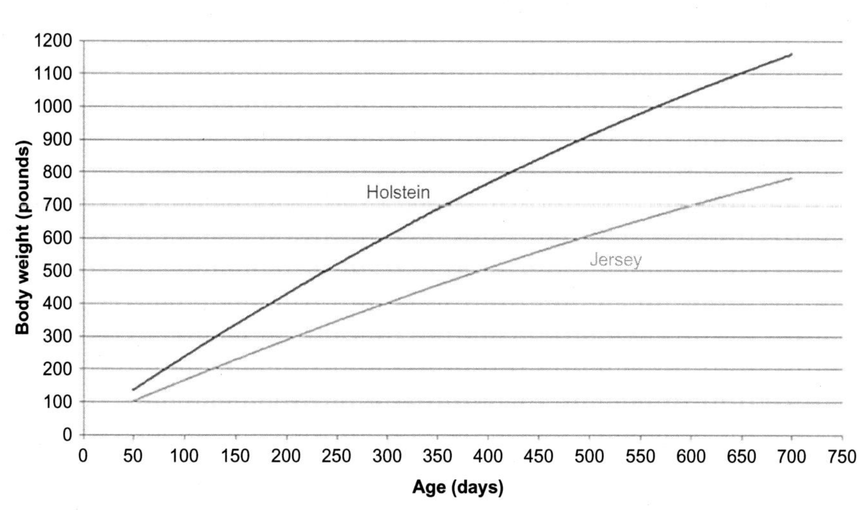

FIG. 6.3

Comparison of Holstein and Jersey cattle for growth traits.

Courtesy, Animal Science Department, Iowa State University.

FIG. 6.4

Hereford cattle representing three maturity levels.

Courtesy, Department of Animal Sciences, University of Wisconsin, Madison.

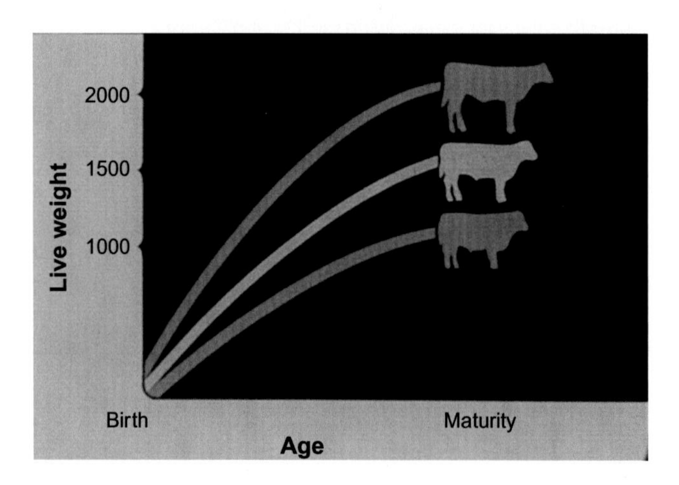

FIG. 6.5

Growth curve examples for Hereford cattle shown in Fig. 6.4.

Courtesy, Department of Animal Sciences, University of Wisconsin, Madison.

representing different maturity levels are shown in Fig. 6.4. The smaller Hereford has growth patterns resulting in a 2-year weight of less than 1000 pounds. The middle Hereford reflects a growth pattern for cattle with a 2-year weight of 1200 pounds. The larger Hereford is typical of cattle with a 2-year weight of 1400 pounds. The cattle in Fig. 6.4 will have three distinct growth curves as shown in Fig. 6.5, indicating that as much variation can exist within breeds as between breeds. This is a very important concept for the use of growth curves in management decisions.

GROWTH CURVES AND PRODUCTION TRAITS

Data from growth curves can also be used for statistical analysis to evaluate the plane of nutrition and environmental interactions when animals are in feedlots, finishing facilities, or even in the early growth phases of production (calves at weaning weight). The growth curve concepts and data can be used to optimize management decisions to predict nutrition needs and daily protein accretion rates. This information can also be used by livestock managers to check the health conditions of the animal by monitoring the normal or less than normal feed intake and weight gains.

When information from growth curves is used to monitor rate of gain and nutritional requirements, some important factors must be kept as constant as possible. For example, the ingestion of food and water just before weighing the animal to obtain growth curve plots must be avoided because the increase in animal weight would be associated with high feed content in the digestive tract and not related to body tissue growth. The same concept applies in the opposite direction if the animals are restricted from normal feed intake for an extended time period before obtaining weights for growth curve plots. This results in lesser contents of feed in the digestive system than when feed intake is normal. Therefore data should be obtained under normal and standard conditions each time weights for normal growth curves are obtained.

STAGES AND PATTERNS OF GROWTH FOR TISSUES AND ORGANS

There is a strong relationship between the stages of growth and body composition of domestic animals. To illustrate the growth patterns, the growth curve can be divided into four stages. When domestic animals grow through the four stages, they follow the law of developmental direction. This law states that the relative tissue deposition for animal growth follows an anterior-posterior pattern for body development. Therefore these growth stages are associated with changes in body shape and conformation and the conformation changes are associated with body composition of the animal.

FIRST GROWTH STAGE

The first growth stage (Fig. 6.6) starts shortly after birth and reflects the early development of the head and neck region as well as the legs. In the first growth stage, these parts are much larger when compared with other parts of the body, such as a more shallow body capacity and lesser developed hindquarters. Organ growth is very active in the first growth stage, muscle and bone growth is active, and fat tissue growth is very limited.

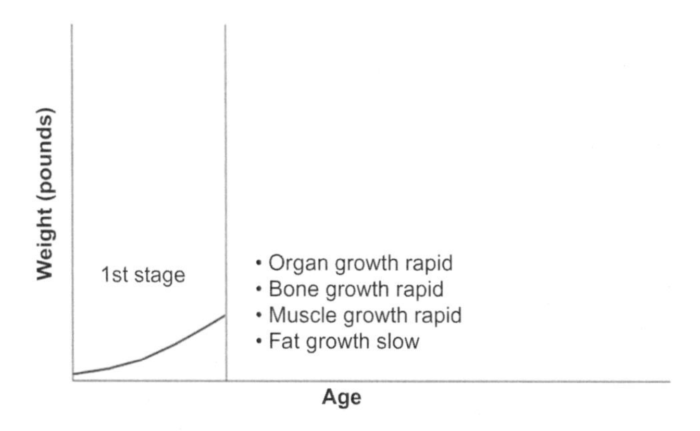

FIG. 6.6

Illustration of the first growth stage on the growth curve.

SECOND GROWTH STAGE

The second growth stage (Fig. 6.7) results in a proportional increase in body length. Therefore body length of the animal is more related with the early growth stages of development than later stages. In the second growth stage, organs are approaching maturity, muscle growth is rapid, and the rate of bone growth starts to slow. Adipose tissue growth begins a significant increase in the later portion of the second stage of growth.

THIRD GROWTH STAGE

The third growth stage (Fig. 6.8) results in a deepening and thickening of the body. Major muscle (protein) and adipose tissue (fat) deposition occurs in this stage and organs reach maturity. Bone growth is nearly completed in the third stage. Under normal growth conditions, this stage starts at approximately 3–4 months in lambs, 4–5 months in pigs, and 10–12 months in beef cattle. These ranges in age can vary depending on the level of nutrition and the genetic potential for growth of muscle and adipose tissue.

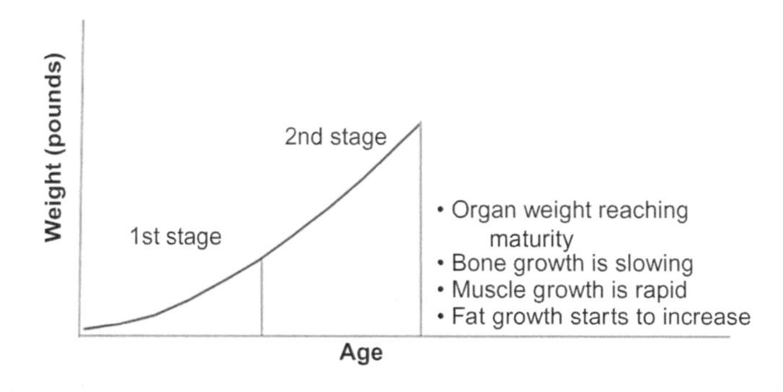

FIG. 6.7

Illustration of the second growth stage on the growth curve.

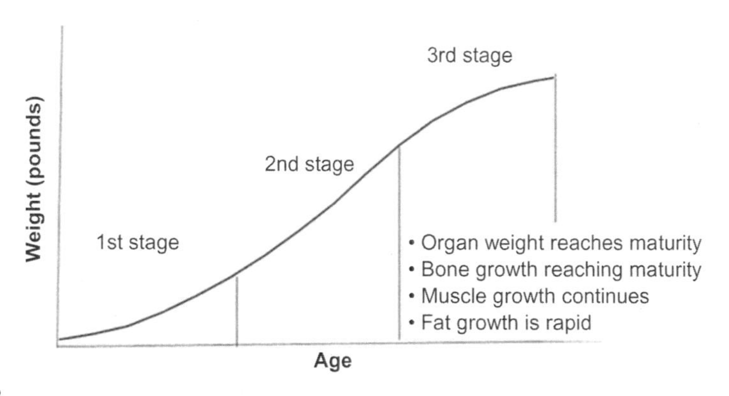

FIG. 6.8

Illustration of the third growth stage on the growth curve.

FOURTH GROWTH STAGE

The fourth stage (Fig. 6.9) involves greater development of the loin and hindquarters, an increase in depth and thickness of the body as muscle growth slows, and the major weight increase is due to fat. Because adipose tissue contains the most calories per unit of weight compared with all other tissues, the increase in bodyweight due to fat deposition results in lower efficiency in rate of gain and a slowing of the growth rate. This explains the flatter portion of the growth curve as an animal reaches maturity.

Knowledge of the four growth stages and the conformation and body composition of the animal at the four growth stages can be helpful when live evaluations of animals take place for carcass traits. A classical summary of the four growth stages is shown in Fig. 6.10.

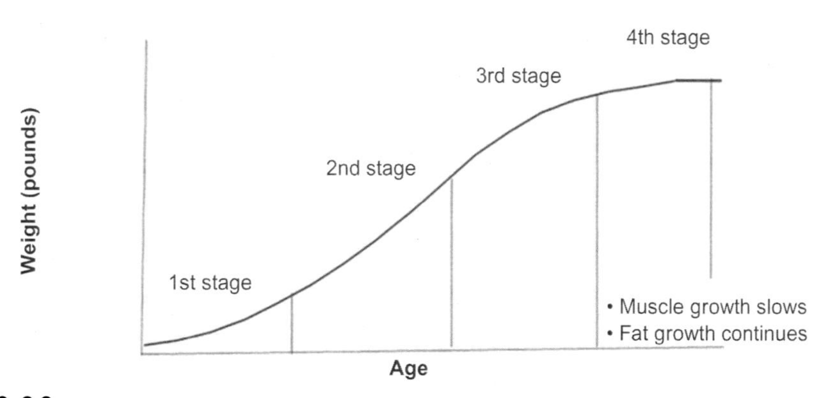

FIG. 6.9

Illustration of the fourth growth stage on the growth curve.

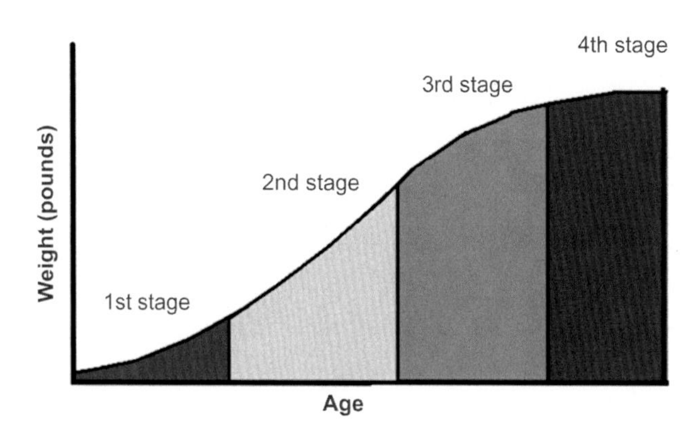

FIG. 6.10

Relationship between the shape of the growth curve and the four growth stages.

CARCASS COMPOSITION TRAITS AND THEIR RELATIONSHIP TO THE SHAPE OF THE GROWTH CURVE

LAMB COMPOSITION AND GROWTH STAGES

Selection for lambs with good muscle traits and reduced fat content continues to be important criteria for the Seedstock producers. It is helpful to understand the normal sequence and magnitude of changes in the carcass composition as lambs grow from a feeder-type lamb to a market weight where lambs are harvested for quality meat consumption. Examples of the sequence are shown in Fig. 6.11. This figure includes the percentage muscle, fat, and bone from lambs representing the medium frame score and weighing 70, 100, and 130 pounds.

The 70-pound lamb has a body composition that is representative of the third growth stage. It has 55.8% lean, 20.4% fat, and 23.8% bone. The 130-pound lamb is an example of the early fourth stage for growth. The carcass composition is 51% lean, 32.3% fat, and 16.7% bone.

Based on the 130-pound weight comparison with the 70-pound lamb, three-fourths of the bone development, one-half of the lean development, and one-third of the fat development had occurred before the lambs weighed 70 pounds. Separable bone decreased 6.1%, separable lean decreased 3.8%, and separable fat increased 9.9% as lambs increased in live weight from 70 to 130 pounds.

CROSS SECTIONS OF LAMBS AT THREE WEIGHTS

The cross sections of lambs at 70, 100, and 130 pounds are shown in Figs. 6.12–6.14. The example of a feeder-type lamb weighing 70 pounds is just starting to have significant deposition of fat in the anterior section of the body. This is shown in the 5th–6th rib cross section. The other cross sections of the 70-pound lamb reflect more limited fat deposition. When the 100-pound lamb is compared at the 6th rib cross

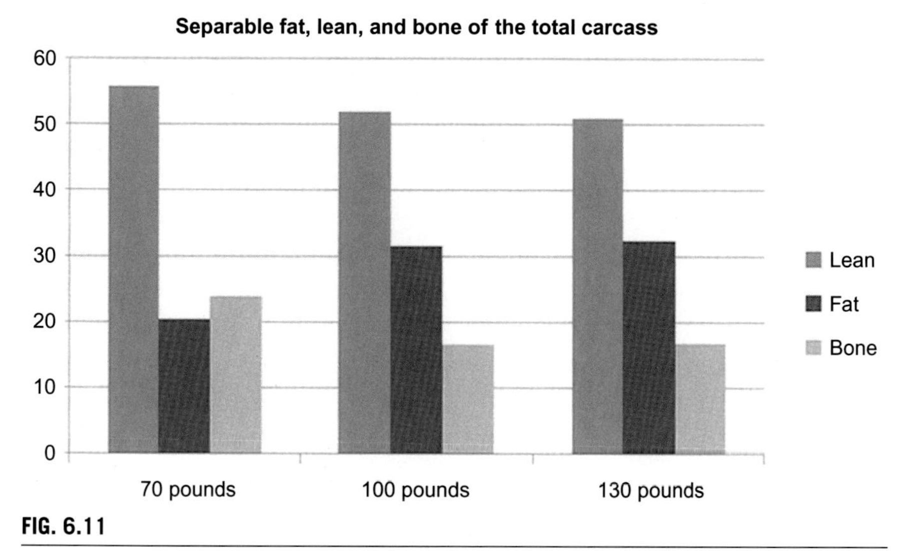

FIG. 6.11

The percentage bone, muscle, and fat from 70-, 100-, and 130-pound lambs.

From G. Rouse and D. Topel. Courtesy, Animal Science Department, Iowa State University.

section, extensive fat is observed in the lower portions of the rib section and the flank section. At this growth stage, seam fat is significantly increased and this results in a widening of the body.

When the cross sections from the 130-pound lamb are observed, the sequence for additional seam fat is evident and major fat is deposited in the lower rib section, lower flank section, and the cross section of the lower hind leg. This is an example of a sequence for fat deposition from the anterior section to the posterior section of the carcass as a lamb grows from 70 to 130 pounds.

RELATIONSHIP OF SWINE GROWTH PATTERNS AND GROWTH STAGES TO PORK CARCASS COMPOSITION

The carcass composition traits (percent skin, bone, muscle, and fat) from pigs weighing 3–300 pounds are presented in Table 6.1. The growth and composition patterns compare genetic lines of fat and meat-type pigs. The percent bone and percent skin decline from birth to 300 pounds in both the meat- and fat-type pigs. The differences in bone and skin percentage are small when the meat- and fat-type pigs are compared.

The highest percent muscle for both the lean- and fat-type pigs is at 100 pounds, but the lean-type pig had 67.1% muscle and the fat-type pig had 60.2% muscle. From 100 to 300 pounds, the percent muscle decreased for both the meat- and fat-type pigs, but the difference in the percent muscle continues to increase at each weight comparison when the two types are compared. At 300 pounds, the carcass from the meat-type pig had 55.6% muscle and the fat-type pig had 40.1% muscle.

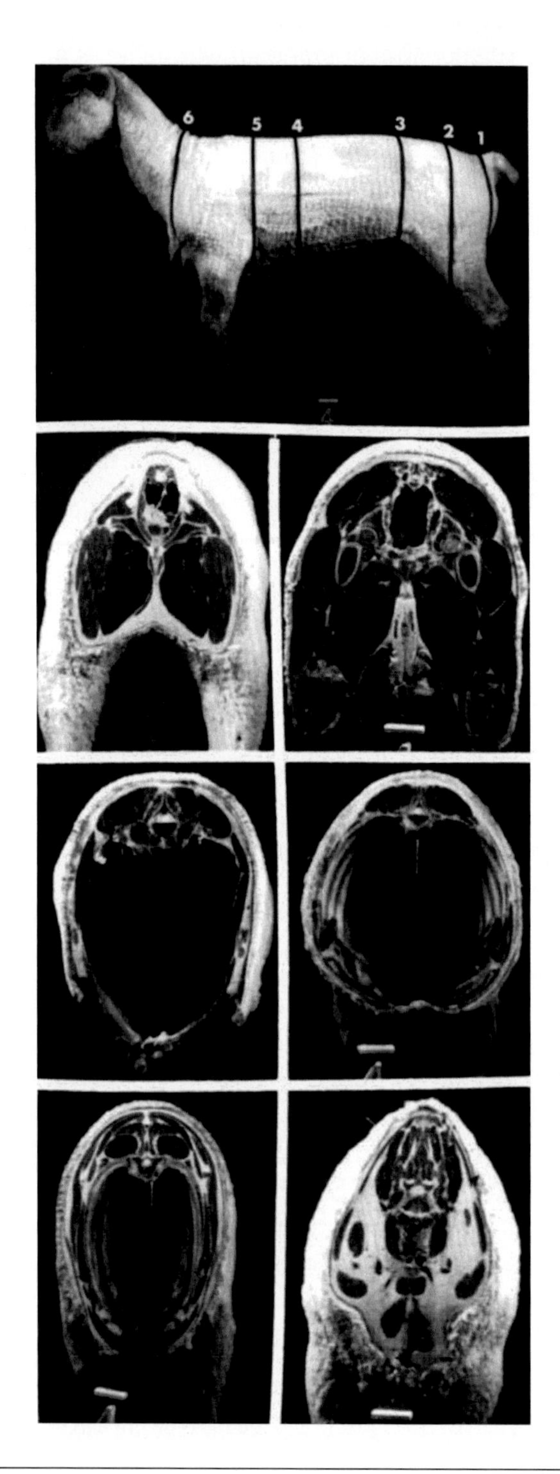

FIG. 6.12

Cross section view of 70-pound lamb. Cross section locations: (1) perpendicular to the junction of the tibia and metatarsus, (2) through the stifle joint, (3) anterior and adjacent to the ilium, (4) between the 12th and 13th rib, (5) between the 5th and 6th rib, (6) anterior to the first rib.

From G. Rouse and D. Topel. Courtesy, Animal Science Department, Iowa State University.

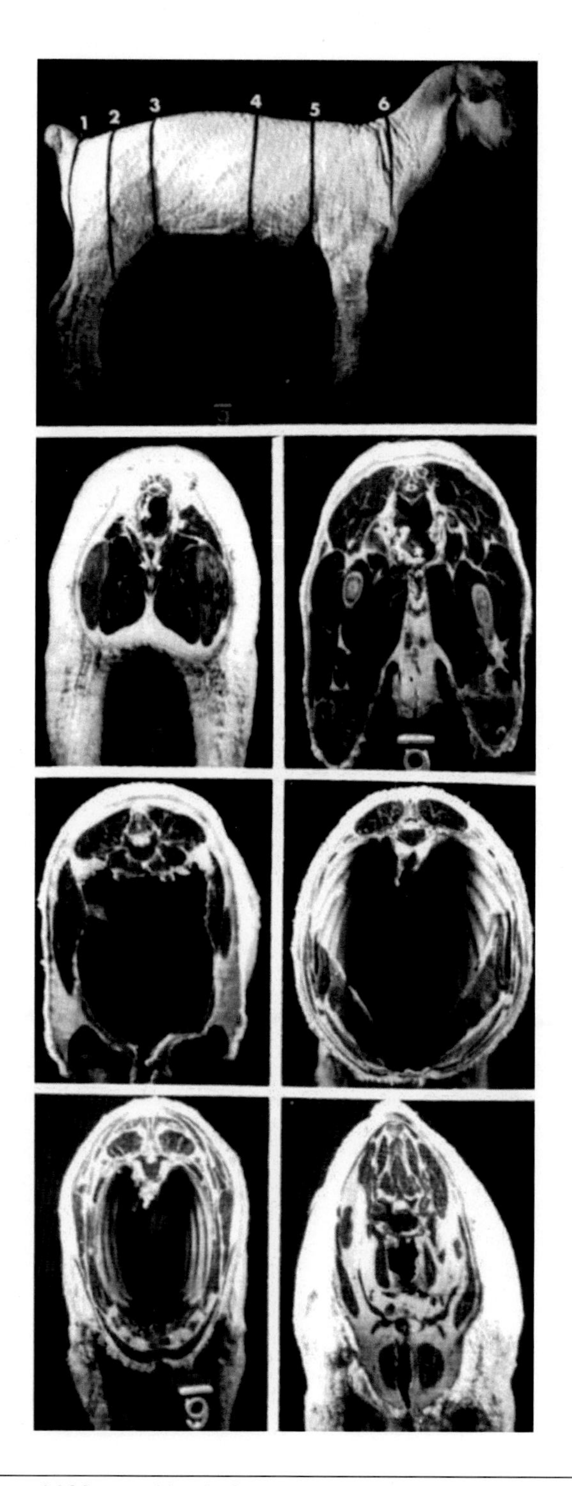

FIG. 6.13

Cross section view of 100-pound lamb. Cross section locations: (1) perpendicular to the junction of the tibia and metatarsus, (2) through the stifle joint, (3) anterior and adjacent to the ilium, (4) between the 12th and 13th rib, (5) between the 5th and 6th rib, (6) anterior to the first rib.

From G. Rouse and D. Topel. Courtesy, Animal Science Department, Iowa State University.

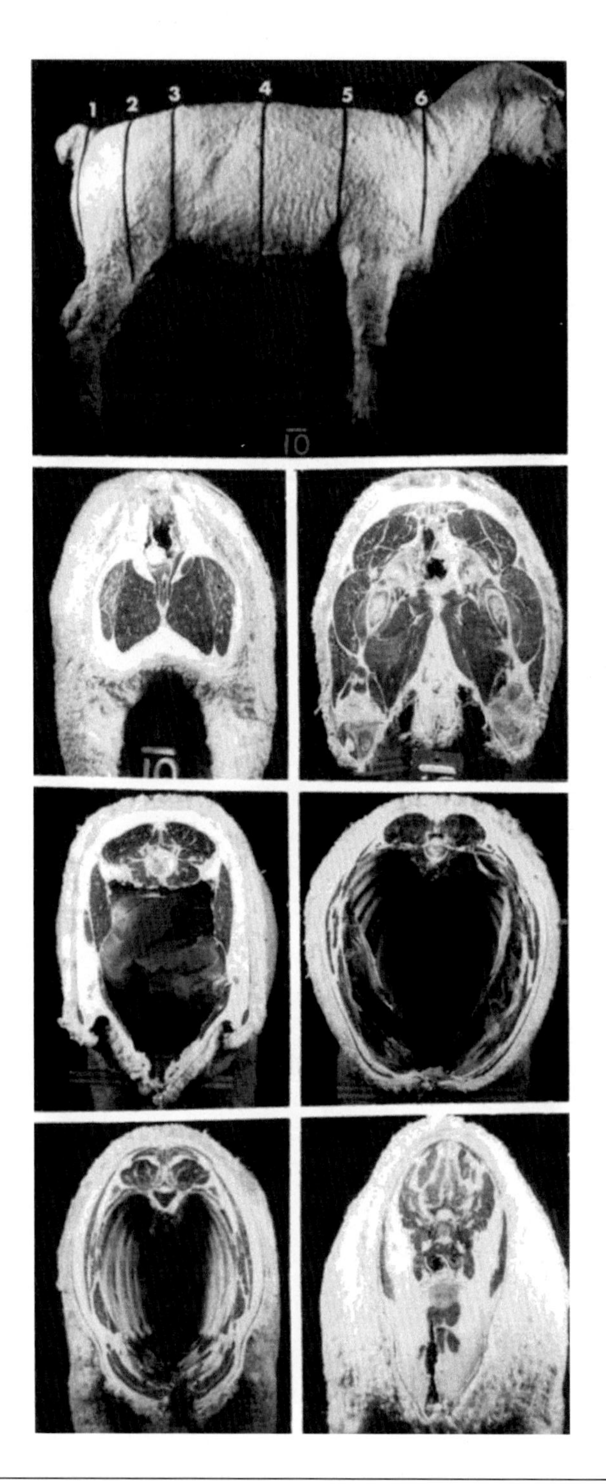

FIG. 6.14

Cross section view of 130-pound lamb. Cross section locations: (1) perpendicular to the junction of the tibia and metatarsus, (2) through the stifle joint, (3) anterior and adjacent to the ilium, (4) between the 12th and 13th rib, (5) between the 5th and 6th rib, (6) anterior to the first rib.

From G. Rouse and D. Topel. Courtesy, Animal Science Department, Iowa State University.

Table 6.1 Relationship between bodyweight and body composition of swine

	Percent muscle		Percent fat		Percent bone		Percent skin	
Live weight (pounds)	**Meat type**	**Fat type**	**Meat type**	**Fat type**	**Meat type**	**Fat type**	**Meat type**	**Fat type**
3	48.6	48.3	0.0	0.0	31.8	31.9	19.3	19.5
50	60.8	59.1	8.9	9.6	18.8	19.1	11.2	11.9
100	67.1	60.2	9.5	13.8	15.9	16.0	9.4	9.9
150	64.0	54.3	14.0	24.4	14.9	14.0	7.0	7.3
200	57.8	46.4	23.8	35.2	12.0	11.5	6.2	6.7
250	56.7	42.3	26.9	41.7	10.6	10.0	5.7	5.9
300	55.6	40.1	29.5	44.2	9.2	9.9	5.6	5.7

The changes in fat percentage in the carcass were similar at 50 pounds, but at 100 pounds, the fat-type pig started to deposit significantly more fat (13.8%) when compared with the meat-type pig (9.5%). The differences in fat deposition continued to widen at each weight comparison from 100 to 300 pounds. At 300 pounds, the fat-type pig had 44.2% carcass fat and the meat-type pig had 29.5% carcass fat. The patterns for fat deposition in pigs for the four growth stages from different genetic lines are shown in Fig. 6.15. Much variation can exist in body composition between different genetic lines of pigs both between breeds and within breeds.

Examples of pigs in weight ranges from 100 to 300 pounds for the meat-type individuals are shown in Fig. 6.16. Examples of the fatter type are shown in Fig. 6.17. Note the changes in body length and the depth of the belly and lower flank of the two

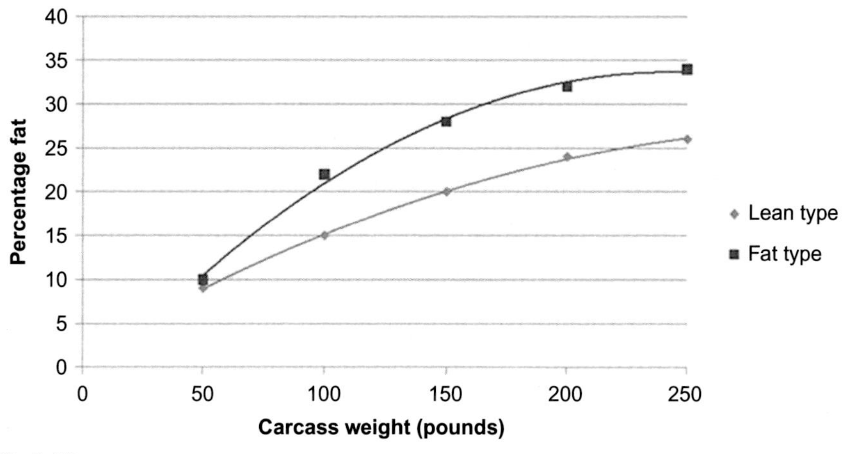

FIG. 6.15

Comparison of percentage fat with carcass weight in meat-type and fat-type pigs.

Courtesy, Animal Science Department, Iowa State University.

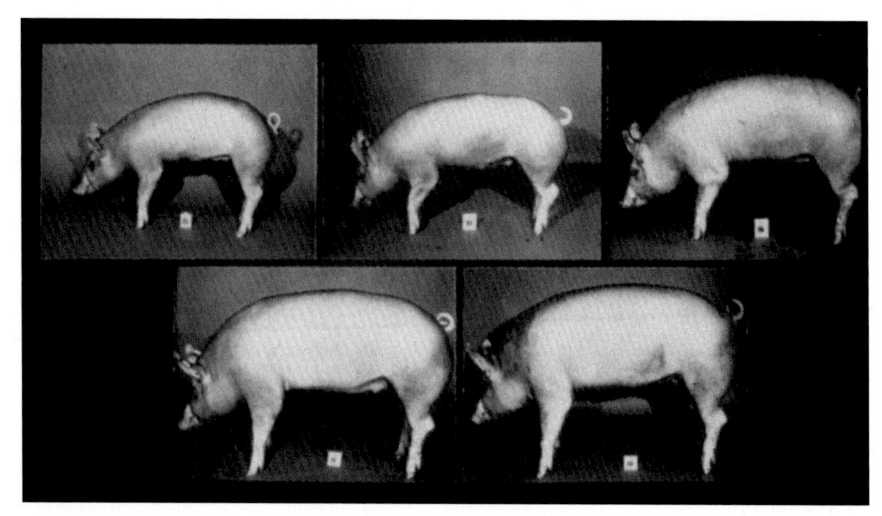

FIG. 6.16

Examples of body type for meat-type pigs from 100 to 300 pounds.

From L. Christian, E. Kline, and D. Topel. Courtesy, Animal Science Department, Iowa State University.

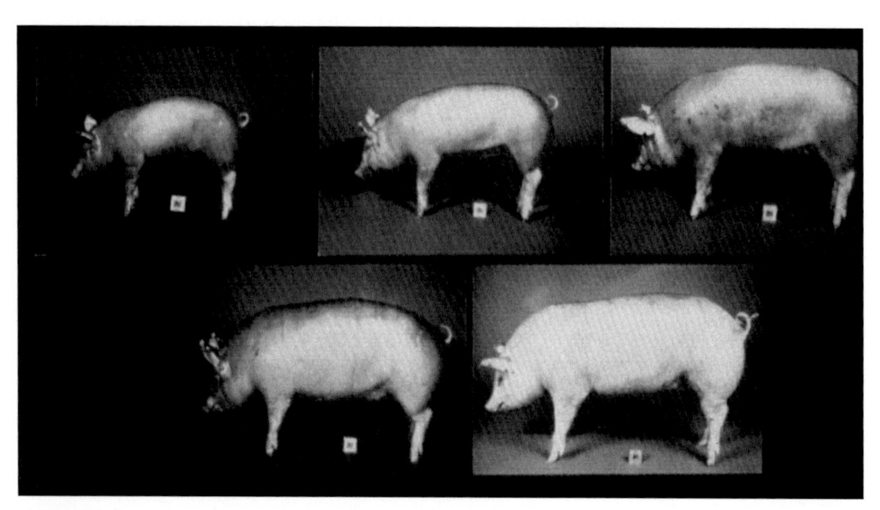

FIG. 6.17

Examples of body type for fat-type pigs from 100 to 300 pounds.

From L. Christian, E. Kline, and D. Topel. Courtesy, Animal Science Department, Iowa State University.

types of pigs as they grow from 100 to 300 pounds. The fat type is deeper in the belly and flank region and the meat-type pigs are longer than the fat type.

Muscle, fat, and bone deposition patterns for the meat-type and fat-type pigs are shown by cross section illustrations in Figs. 6.18 and 6.19. Each growth stage is illustrated. The cross section illustrations in Figs. 6.18 and 6.19 reflect the physical changes in bone, muscle, and fat deposition for a meat-type and fat-type pig at different growth stages. The cross section illustrations represent pigs with similar body composition as those presented in Table 6.1. Note the deepening of the body as the pig matures, and the deepening is due to extensive fat deposition in the lower flank, belly, and lower rib section. Also, the backfat and intermuscular fat are greatly increased as the pig reaches the last two growth stages. Note these changes in Figs. 6.18 and 6.19.

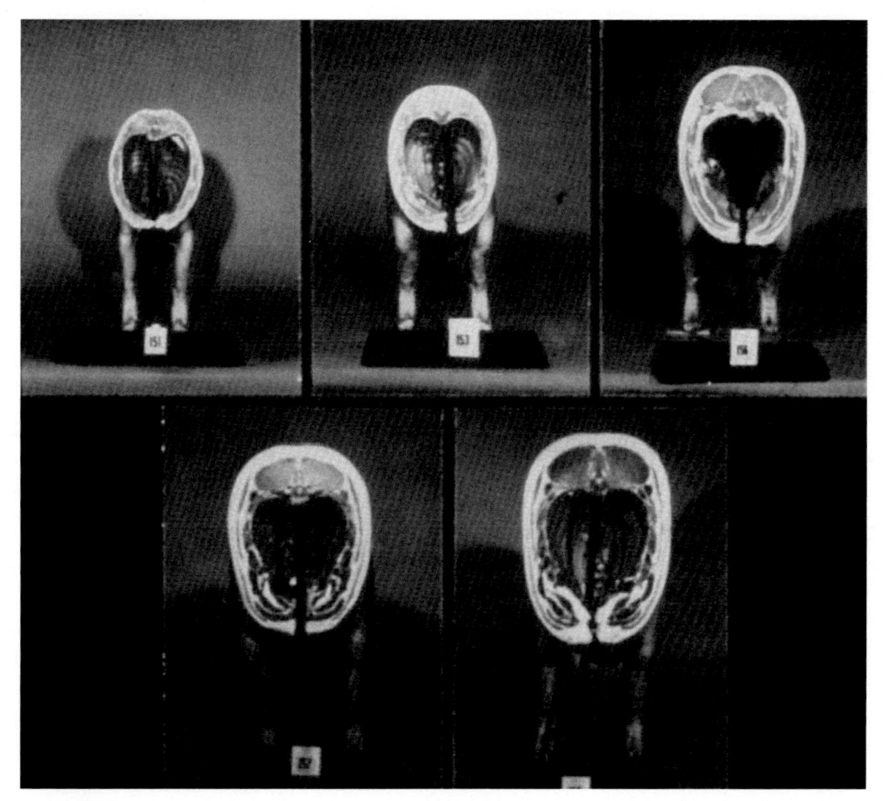

FIG. 6.18

Examples of physical traits of muscle, fat, and bone in meat-type pigs.

From L. Christian, E. Kline, and D. Topel. Courtesy, Animal Science Department, Iowa State University.

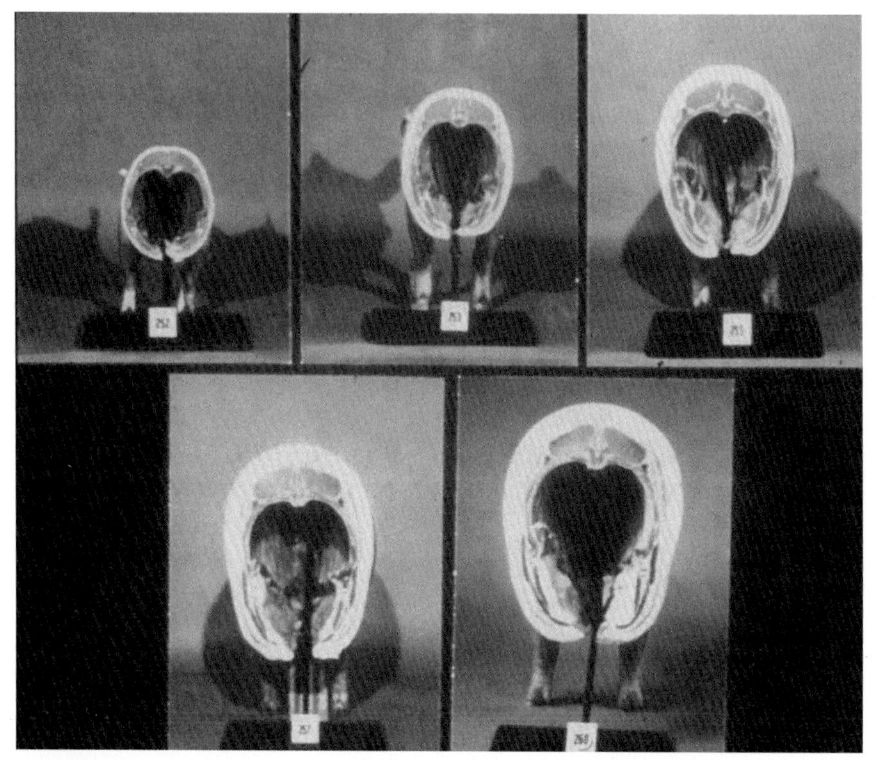

FIG. 6.19

Examples of physical traits of muscle, fat, and bone in fatter-type pigs.

From L. Christian, E. Kline, and D. Topel. Courtesy, Animal Science Department, Iowa State University.

GROWTH PATTERNS FOR SUBCUTANEOUS FAT DEPOSITION IN CATTLE

A comparison of Angus steers and heifers with similar genetic backgrounds is shown in Figs. 6.20–6.22. The fat contours represent patterns for Angus cattle in the third growth stage and the beginning of the fourth growth stage. The darker patterns represent less than 0.4 in. of subcutaneous fat. The lighter patterns represent more than 0.80 in. of subcutaneous fat, and the middle pattern ranges from 0.40 to 0.80 in. of subcutaneous fat. Heifers' fat patterns in Fig. 6.20 reflect greater fat in the center of the chuck, lower flank, and the lower portion of the loin. Also, more brisket and rump fat is evident in the heifer examples when compared with steers. The 925-pound example (Fig. 6.21) reflects a similar fat pattern for the steer carcass as exists for the 850-pound Angus heifer. The 925-pound Angus heifer has a continuous pattern of fat that is greater than 0.8 in. from the lower flank and rump region through the loin and rib region, and it also joins the chuck portion of the carcass. The 925-pound

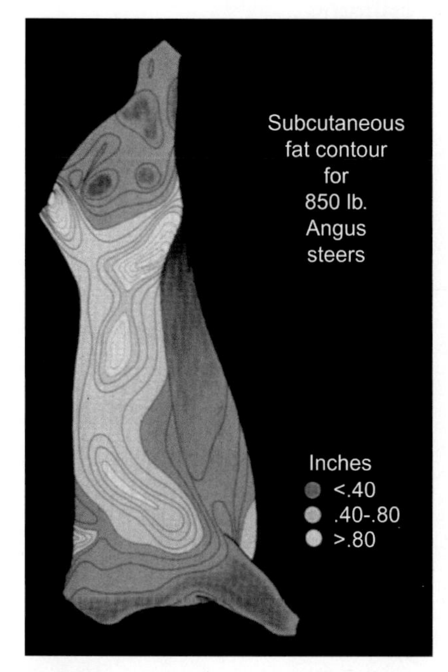

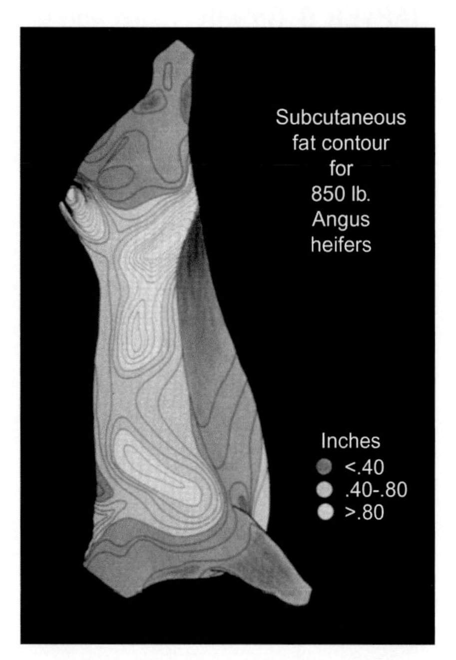

FIG. 6.20

Subcutaneous fat contours and patterns for an 850-pound Angus steer and heifer.

Courtesy, Department of Animal Sciences, University of Wisconsin, Madison.

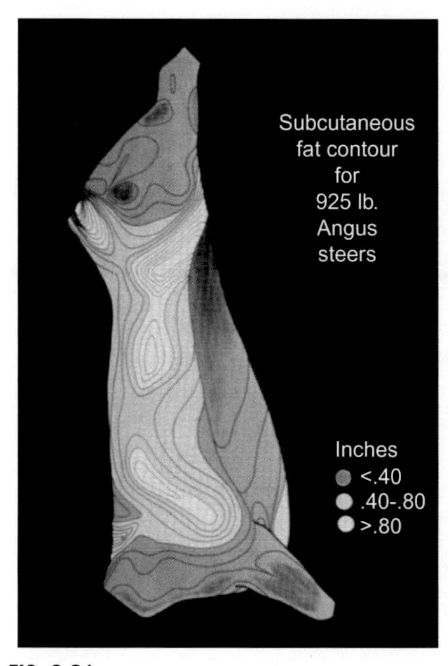

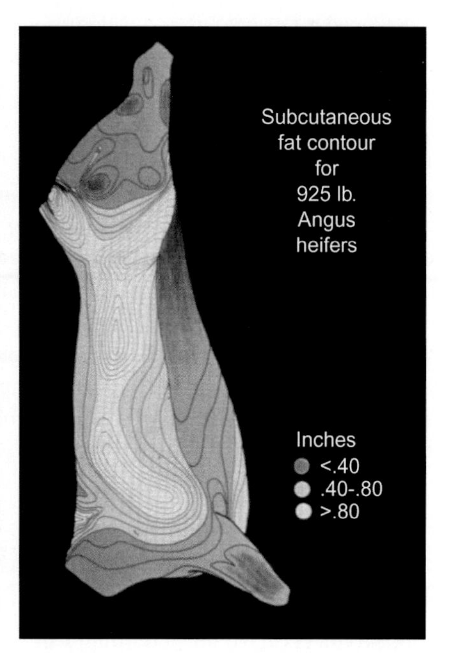

FIG. 6.21

Subcutaneous fat contours and patterns for a 925-pound Angus steer and heifer.

Courtesy, Department of Animal Sciences, University of Wisconsin, Madison.

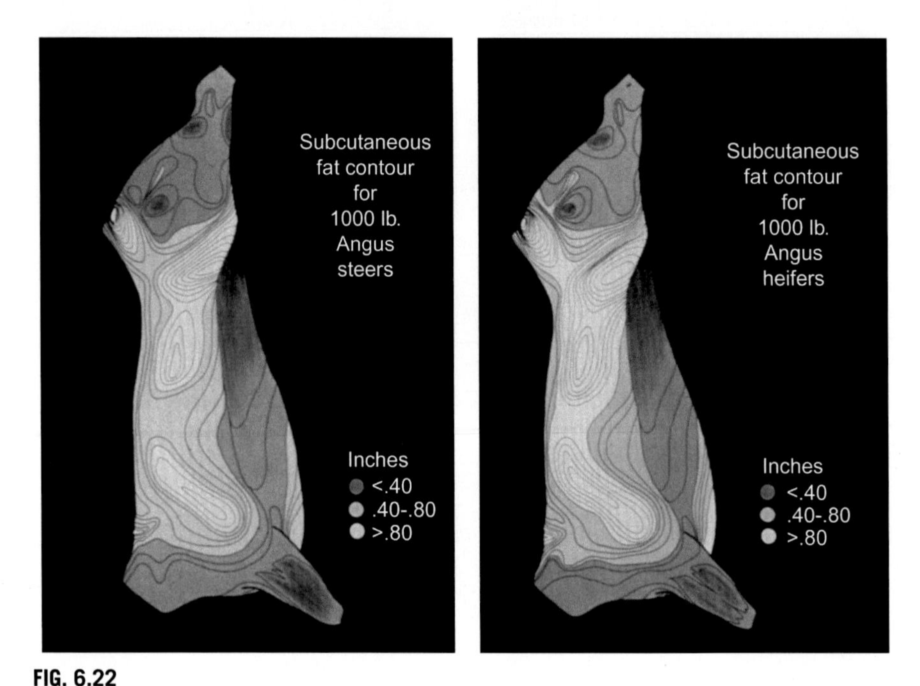

FIG. 6.22

Subcutaneous fat contours and patterns for a 1000-pound Angus steer and heifer.

Courtesy, Department of Animal Sciences, University of Wisconsin, Madison.

heifer is in the early fourth stage of development and the 925-pound steer example is in the third growth stage. The 1000-pound example (Fig. 6.22) reflects an expansion of the fat patterns for both the heifer and steer carcasses, but the steer carcass still does not have a continuous pattern of subcutaneous fat that is greater than 0.8 in. from the flank, rump, loin, rib, and chuck regions of the carcass as is present for the 1000-pound heifer carcass. These fat deposition patterns reflect the genetic differences for fat deposition due to the sex of the animal for cattle. Sheep have similar fat patterns at the third and fourth growth stage for the ewe and wether lambs, but pigs are different. Gilts have less fat than barrows when fat patterns are compared at the third and fourth growth stages.

GROWTH PATTERNS FOR BACKFAT THICKNESS, MUSCLING, AND MARBLING IN TWO GENETIC LINES OF DUROCS

Pork producers have an opportunity to genetically select pigs to be sold on an incentive basis for increased muscling, less fat, and in some markets, more marbling. Therefore knowledge of growth patterns for these traits is important for selection and marketing. Pork producers can use data obtained from growth curves for genetic selection indices and improve carcass traits that have economic importance. Examples of two genetic lines within the Duroc breed will be presented for genetic comparisons in Fig. 6.23. Genetic lines from pigs representing two different levels of carcass fat were compared. Differences in bodyweight for the two genetic lines are shown in Fig. 6.24. At 100–180 days of age, the Duroc line (reflecting the fatter type of pigs) averaged about 10 pounds heavier than the more muscular pigs, but the average daily gain was about the same for the two types of Durocs. At 180 days, both genetic types of Durocs weighed approximately 260 pounds.

BACKFAT THICKNESS

The backfat thickness values for the two Duroc lines are shown in Fig. 6.25. At 160 pounds live weight, the Durocs from the fatter genetic base had 0.7 in. of backfat at the 10th rib, and the more muscular genetic line of Durocs had 0.57 in. of backfat at the 10th rib.

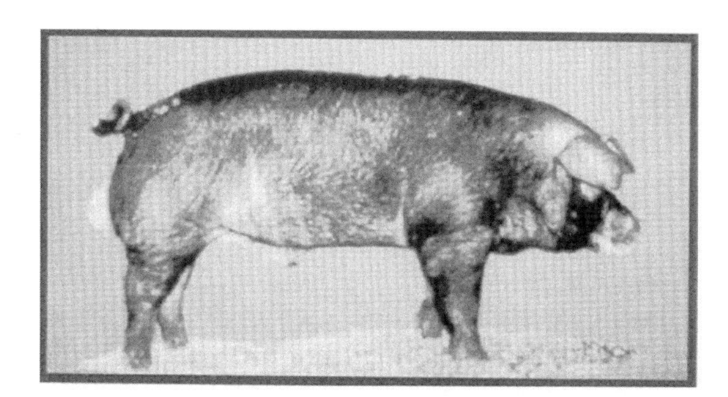

Fatter type

Leaner type

FIG. 6.23

Example of two genetic lines of the Duroc breed used for the growth comparisons in Figs. 6.24–6.27.

From T. Bass and C. Schwab. Courtesy, Animal Science Department, Iowa State University.

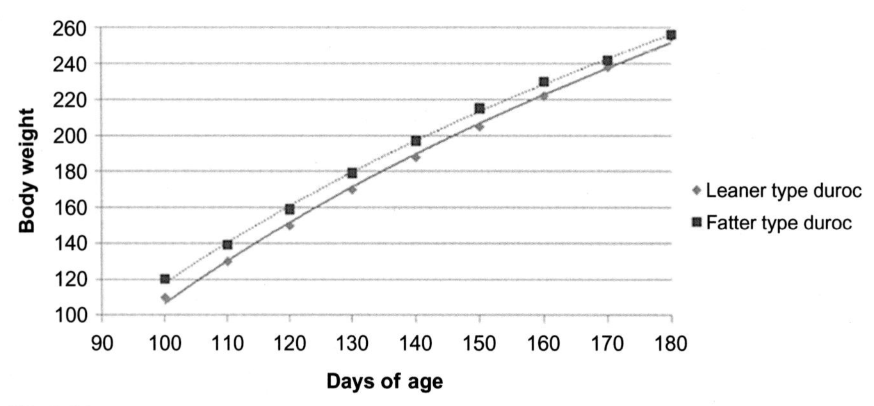

FIG. 6.24

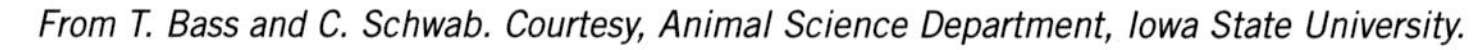

Growth patterns of bodyweight change in pounds for purebred Durocs sired by boars from two genetic lines.

From T. Bass and C. Schwab. Courtesy, Animal Science Department, Iowa State University.

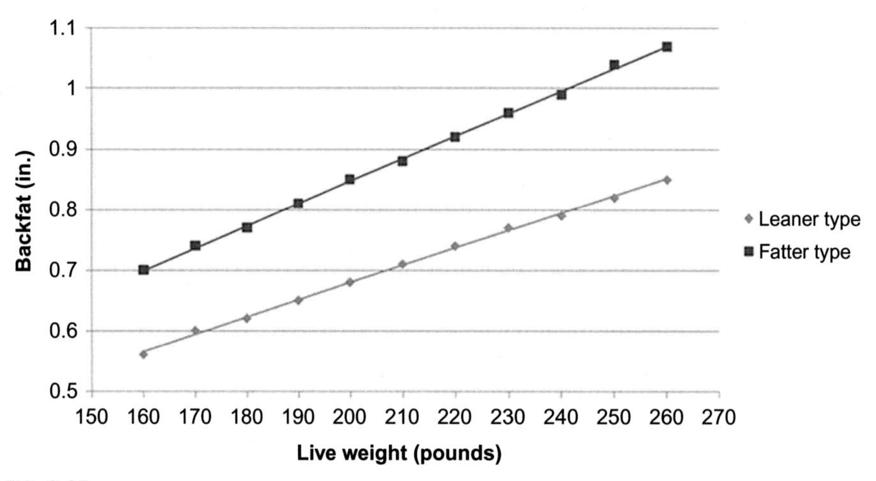

FIG. 6.25

Growth patterns for the tenth-rib backfat from purebred Durocs sired by boars from two genetic lines.

From T. Bass and C. Schwab. Courtesy, Animal Science Department, Iowa State University.

Growth from 160 to approximately 260 pounds resulted in a larger and consistent backfat increase for each 10-pound live weight increase for the fatter genetic line Durocs when compared with the more muscular Durocs. These trends are shown in Fig. 6.25. The margin of difference starts to widen at 210 pounds, and at 260 pounds, the fatter genetic line Durocs had 1.08 in. of backfat and the more muscular Duroc lines had 0.77 in. of backfat at the 10th rib.

LOIN MUSCLE AREA

Growth patterns for loin muscle area of two genetic types of Durocs are shown in Fig. 6.26. The patterns are similar for both genetic types when compared from 160 to 260 pounds. The more muscular Duroc lines, however, had a larger loin muscle area at each weight comparison. At 160 pounds, the difference was 0.4 in.^{2}, and at 260 pounds, the difference was 0.9 in.^{2}. The two Duroc lines continued to increase the loin muscle area even at 260 pounds. Therefore the two Duroc lines were still in the third growth phase, but the more muscular Durocs reflected greater muscle deposition traits at a heavier live weight. This indicates great genetic improvement within the Duroc breed by intensive selection for increased muscle.

INTRAMUSCULAR FAT (MARBLING) CHANGES

Changes in marbling in the longissimus muscle (loin eye muscle) for the two genetic lines of Durocs are shown in Fig. 6.27. At 160 pounds, Durocs representing the fatter genetic line had more marbling in the loin muscle than the more muscular-type Durocs. This difference continued to increase as growth occurred to 260 pounds (3.8% vs 3.6% at 160 pounds and 4.8% vs 4.2% at 260 pounds).

When the two genetic lines of Durocs are compared for growth patterns related to carcass composition, much progress was achieved by intensive selection by the Duroc breeders for reduced fat and increased muscle. These changes have altered the growth patterns, reduced carcass fat content, and increased carcass muscle percentage. The changes in growth patterns also resulted in a reduction of intramuscular fat (marbling) and eating quality traits such as flavor.

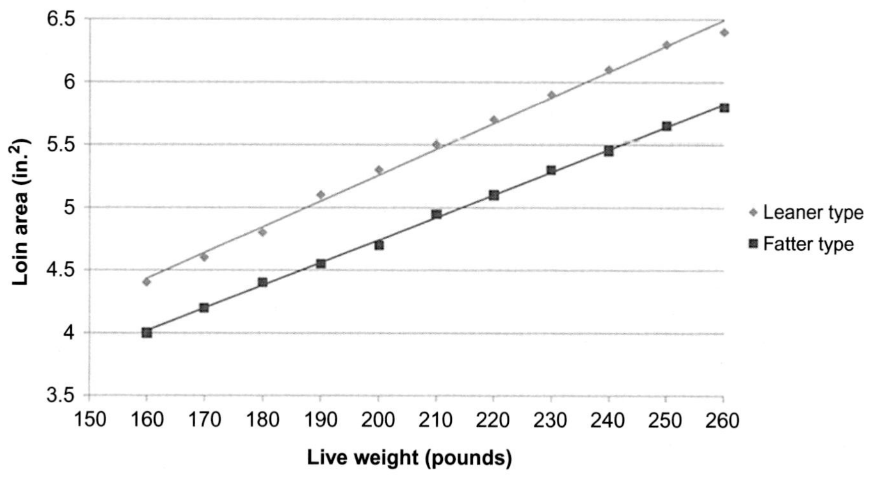

FIG. 6.26

Growth patterns for the loin eye area for purebred Durocs sired by boars from two genetic lines.

From T. Bass and C. Schwab. Courtesy, Animal Science Department, Iowa State University.

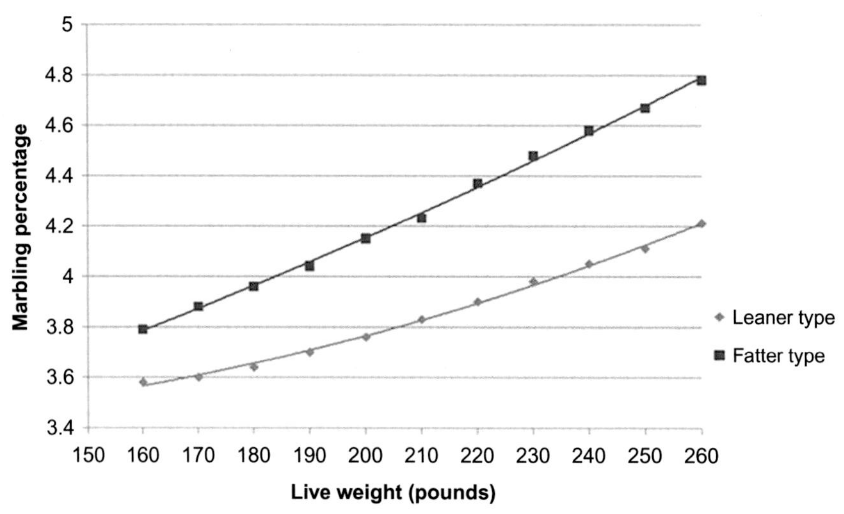

FIG. 6.27

Growth patterns for intramuscular fat (marbling) for purebred Durocs sired by boars from two genetic lines.

From T. Bass and C. Schwab. Courtesy, Animal Science Department, Iowa State University.

This growth and composition information indicates that through good genetic selection, Seedstock producers can design and produce pigs with specific carcass traits (highly marbled meat, for example) for niche markets and receive a higher price. An understanding of growth patterns for animals is required for successful genetic selection of market traits (carcass composition) at different live weights.

The results from the comparison of the fatter type and more muscular Duroc lines also indicate that it is feasible to use genetic archives developed by purebred producers as well as companies that provide semen to the industry for the development of market pigs for specific or niche domestic or international markets.

The concepts illustrated with the two Duroc lines can also be applied to other species, including beef cattle and sheep.

GROWTH CURVES AND MARBLING ESTIMATES FOR BEEF CATTLE

With ultrasound technology, marbling can be determined in bulls and heifers as they grow through the growth stages starting with ultrasound marbling values at weaning weights. Values for Angus bulls from 220 to 400 days are shown in Fig. 6.28. This marbling growth curve is representative of typical Angus bulls, but the rate of intramuscular fat deposition can vary extensively when different sires are compared in progeny tests. Genetic differences from progeny tests for intramuscular fat (marbling) between Angus bulls are shown in Fig. 6.29. At 220 days, the Angus sire line

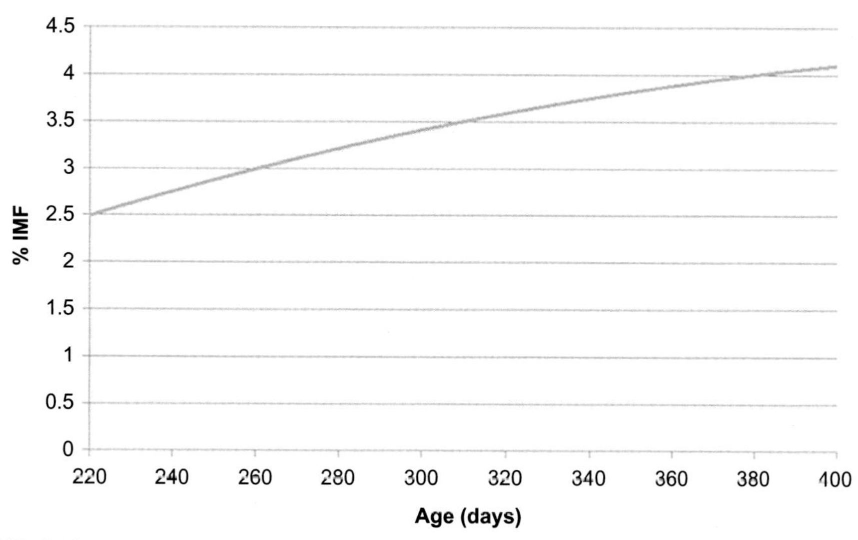

FIG. 6.28

Percentage intramuscular fat in Angus bull calves from weaning to yearling age.

From G. Rouse. Courtesy, Animal Science Department, Iowa State University.

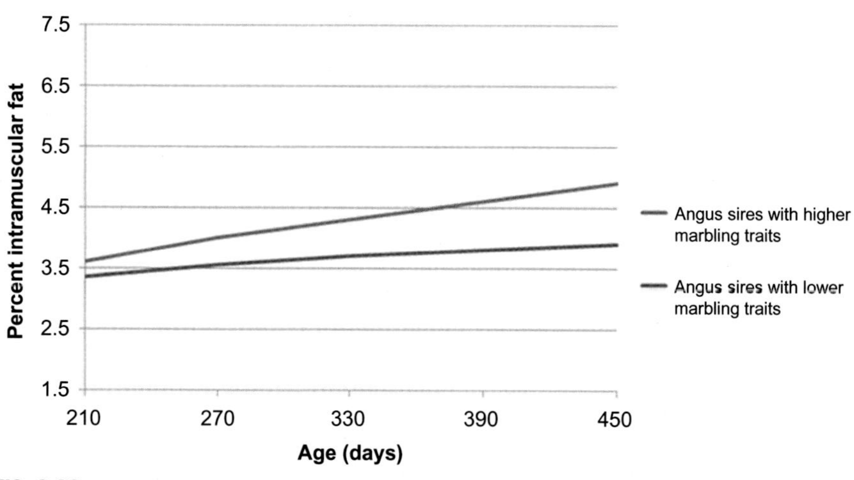

FIG. 6.29

Rate of intramuscular fat between progeny of different Angus sires.

From G. Rouse. Courtesy, Animal Science Department, Iowa State University.

with greater genetic potential for marbling deposition had 3.6% intramuscular fat and the Angus sire line with less genetic potential for intramuscular fat had 3.3%. When the two genetic lines are compared at 360 days, the sire line for greater genetic potential for intramuscular fat has 4.5% and sire lines with less genetic potential for intramuscular fat increased only to about 3.6%. These sire lines show major

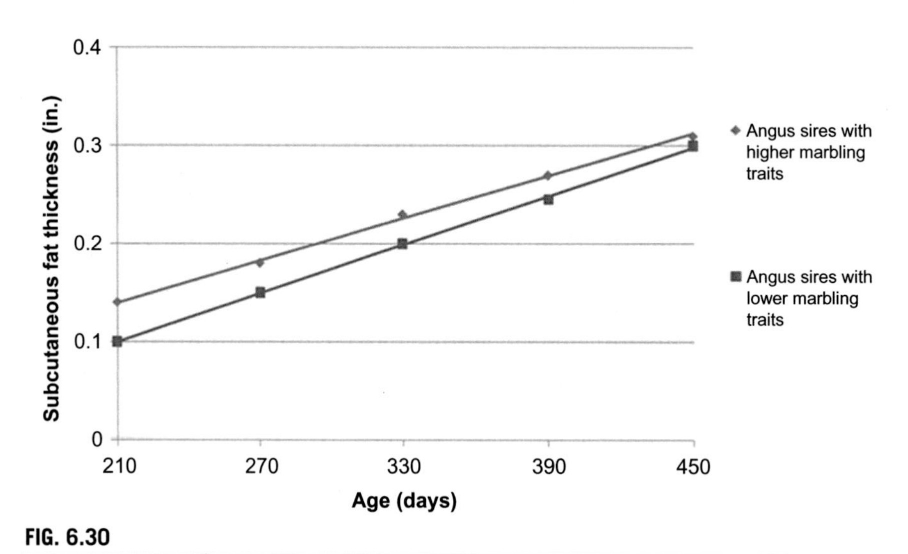

FIG. 6.30

Rate of subcutaneous fat thickness deposition at the 12th rib for Angus cattle from 210 to 450 days of age.

From G. Rouse. Courtesy, Animal Science Department, Iowa State University.

differences in intramuscular fat deposition rate. The subcutaneous fat deposition rate, however, of the two genetic sire lines at 450 days was similar (Fig. 6.30). This suggests that the relationship between subcutaneous fat and intramuscular fat in beef cattle is very low.

GROWTH CURVE CONCEPTS FOR PRODUCTION AND MARKETING OF CATTLE

PRODUCTION CONCEPTS

Managing production concepts during the growth process for beef cattle requires a long-term plan that has an important link between production and all phases of the food chain. From the production phase, ruminants have an advantage over nonruminants as they can convert roughage feeds and by-products into highly digestible food with good protein quality for the human diet. Utilization of roughage in the ration should be a major part of the Seedstock and Commercial Cow-Calf operations to improve profits. Nearly one-half of the land in the United States and two-thirds of the land in the world is best suited for production of forage-type feed. When evaluating cow-calf programs, cows are seldom fed grain diets because they are too costly. Cows are normally grazing on land that best supports forage-type systems and not tillable land for grain crops. Often, cow-calf programs graze the land that receives limited rain fall but enough moisture to support forage systems. The utilization of quality forage for cows results in a good nutritional program that is also economical.

The most important nutritional contribution is to make sure that the grazing cattle have quality nutritional supplements to foster good maintenance levels for the cows. Good forage systems also protect the soil from erosion and are beneficial to the environment. A high roughage diet results in enough energy to maintain the breeding stock, and therefore they are good diets for the Seedstock and Cow-Calf operators to improve efficiency for protein deposition in the live animal rather than fat deposition. Therefore beef cattle can be a major source of quality protein foods around the world.

When cattle enter the feedlots, the managers switch from a roughage-base ration for growth of feeder calves to a high-energy ration designed to fatten the cattle to reach the Choice grade. This process requires a good management and nutrition plan to manage the muscle and fat deposition stages illustrated in the growth curves (Figs. 6.8 and 6.9). The growth curves in Figs. 6.8 and 6.9 could be representative of a 1250-pound market steer.

APPLICATION OF GROWTH CURVE CONCEPTS FOR MANAGEMENT OF FAT DEPOSITION

The 1250-pound market steer used as an example for the growth curves cited previously will grow slowly initially and then in the next stage shown in the growth curve, the steer grows more rapidly. At this stage, the steer is putting on a lot of muscle and bone. Muscle contains a large amount of water and protein and only a small amount of fat at this growth stage. When the muscle and bone growth starts to decline, the fat growth starts to increase and can increase at a rapid rate depending on the genetic potential of cattle for fat deposition. Much variation exists between beef cattle in their ability to deposit fat in large quantities at the beginning of the 4th growth phase. It is at this stage of growth that management of fat deposition becomes important to make sure feed efficiency is still adequate, marbling is at a level to grade USDA Choice or higher, and subcutaneous fat is not excessive, resulting in better profits for the feedlot operations.

To have a better understanding on how to manage fat deposition at the 3rd and 4th growth stages, a review of the types of fat deposition in cattle may be helpful. Fat deposition can be divided into four separate depots. Three of the four can be considered waste fat. The three listed as waste fat for retail marketing of beef are internal fat that surrounds the organs, seam fat that is located between muscles, and subcutaneous fat that is located under the hide of cattle. The 4th type of fat is marbling (intramuscular fat). Marbling is related to the flavor of beef steaks and roasts and it is often referred to as taste fat.

For many years, feedlot managers had no objective methods to determine how much marbling was present in the muscle of the cattle fed in the feedlots and they would often feed the cattle so excess waste fat was deposited before the cattle were sent to the packer for the harvest process. The cattle feeder associated a relationship between subcutaneous fat and marbling (Fig. 6.31). This practice resulted in the marketing of too many cattle with excess waste fat and poor feed efficiency (see Example A in Fig. 6.31).

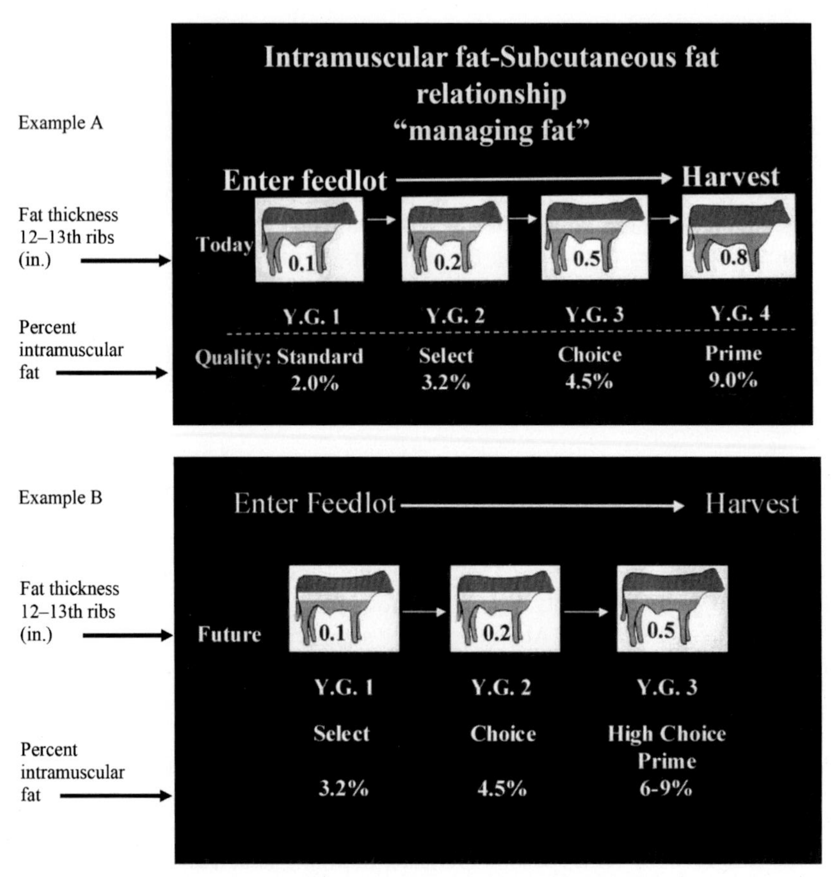

FIG. 6.31

An example of intramuscular and subcutaneous fat relationships when cattle enter the feedlot and when they are harvested. Example A is for typical programs and Example B is for more modern programs.

Courtesy, Dr. Gene Rouse, Iowa State University.

When cattle in the feedlot reach a weight related to the 4th growth stage, it takes 10 pounds of waste fat deposition for the deposition of one pound of taste fat. This results in poor feed efficiency, and depending on the genetic potential of the feeder cattle for marbling deposition, the feeder cattle may not continue to deposit more marbling even though excess subcutaneous and other waste fat continues to be deposited. To overcome excess feeding of cattle to reach the USDA Choice grade, an adequate management and marketing plan is needed. With the use of growth curve data, ultrasound technology, and cooperation between all segments of the beef industry, it is possible to market beef cattle from the feedlot that have good taste fat (Choice and Prime carcasses) and a small amount of waste fat (see Fig. 6.31, Example B).

BACKGROUND INFORMATION NEEDED FOR AN INDUSTRY MANAGEMENT PLAN FOR BEEF CATTLE

The beef industry has developed adequate plans that will provide guidance to all segments of the beef industry to implement a plan to produce and market cattle with excellent meat quality (good taste fat) and without excess waste fat. These plans are based on the following concepts:

1. The genetic correlation or relationship between subcutaneous fat and intramuscular fat is low in beef cattle.
2. A large number of yearling bulls and replacement heifers in the Seedstock industry must be evaluated with real-time ultrasound for body composition traits that include the rib-eye area, subcutaneous fat thickness, and the degree of intramuscular fat.
3. Ultrasound measurements for a large number of bulls and replacement heifers along with carcass data (e.g., waste fat and taste fat measurements) on progeny need to be converted into expected progeny difference (EPD) values by the Seedstock industry. This becomes a part of the genetic base required to implement the plan.
4. The management of the packing industry must obtain the carcass information from the cattle slaughtered that were part of the production plan. The carcass data are then sent to the Seedstock producer, the cow-calf operators, and the owners of the feedlot cattle. The EPD values are also developed.

ESTABLISH A GENETIC BASE AT DIFFERENT GROWTH STAGES FOR SELECTED CARCASS TRAITS

Until recently, the only way to obtain data for carcass traits as cattle grow from weaning to slaughter (harvest) weights was to harvest the cattle at each growth stage. This was an expensive process and, therefore a small number of cattle were often used for each growth stage. With the development of quality real-time ultrasound technology, accurate values for intramuscular fat (marbling) can be determined on the same animal at all of the growth stages. Therefore this data can be used for genetic selection of intramuscular fat in beef cattle and improve the amount of taste fat for quality beef cuts.

SERIAL ULTRASOUND SCANNING

To obtain a good genetic selection base for intramuscular fat, subcutaneous fat thickness, and rib-eye area, serial ultrasound values need to be obtained at each growth stage, shown in Figs. 6.8 and 6.9. Iowa State University researchers under the leadership of Dr. Gene Rouse and Dr. Doyle Wilson developed experiments to establish a database to determine EPD values for carcass traits that included intramuscular fat, rib-eye area, and subcutaneous fat thickness. The bulls were scanned with a real-time ultrasound machine every 28 days for intramuscular fat, subcutaneous fat thickness,

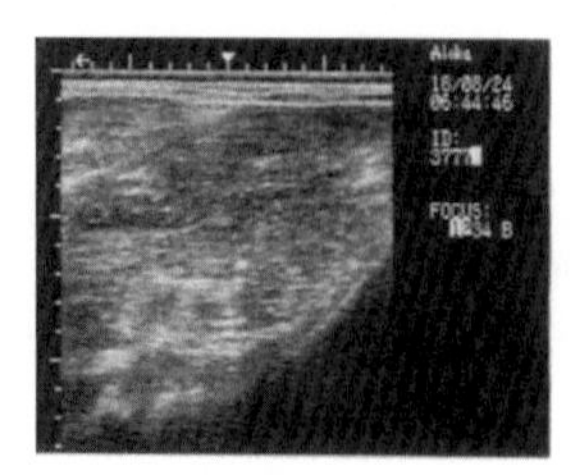

Image 1

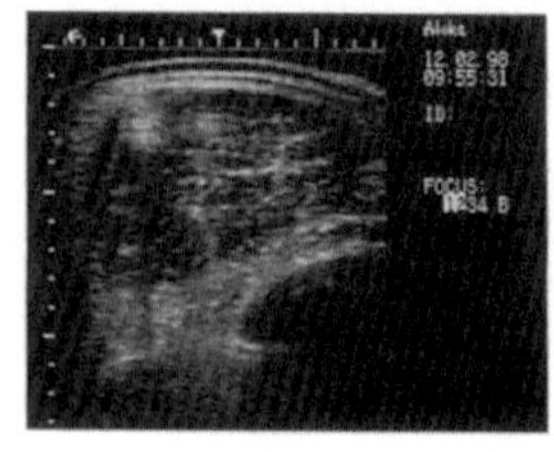

Image 2

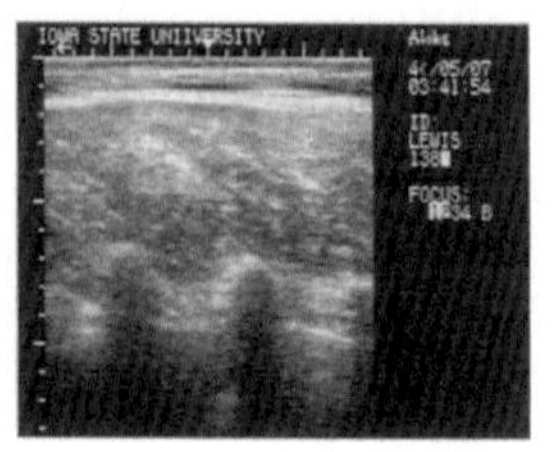

Image 3

FIG. 6.32

Images obtained from ultrasound scans of cattle. Image 1 is from the rump section, Image 2 is a cross section at the 12th and 13th rib, and Image 3 is a longitudinal scan of the rib eye showing intramuscular fat.

Courtesy, Dr. Gene Rouse, Iowa State University.

and rib-eye area. Live weight was also obtained every 28 days. Examples of the ultrasound images that were obtained are shown later. Image 1 of Fig. 6.32 shows the ultrasound scan at the rump area of the steer and Image 2 shows the ultrasound scan at the 12th–13th rib region for rib-eye area and subcutaneous fat thickness. Image 3 of Fig. 6.32 shows the intramuscular fat in the rib-eye muscle at the 12th–13th rib sections.

GROWTH CURVES OBTAINED FOR THE PRODUCTION AND CARCASS TRAITS

Fig. 6.33 indicates the increase in weight of the cattle used in the ultrasound, growth curve study. The cattle were in the study from 600 to 1175 pounds or for 1 year. The rib-eye area, subcutaneous fat at the 12th–13th rib, and intramuscular fat were determined at each live weight at which the cattle were weighed (every 28 days). The rate of live weight gain from 320 days was 3.5 pounds per day, and then as shown in Fig. 6.34, the average daily gain declined from 320 to 400 days. At this stage of growth, an increase in fat is starting to be deposited, resulting in a decline in the average daily gain.

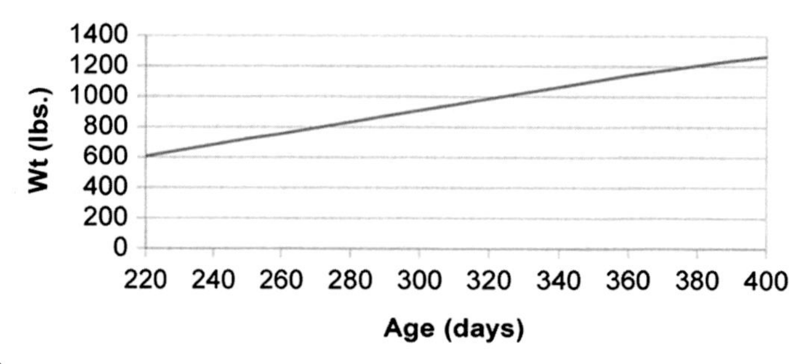

FIG. 6.33

The live weight gains of cattle from 600 to 1175 pounds.

Courtesy, Dr. Gene Rouse, Iowa State University.

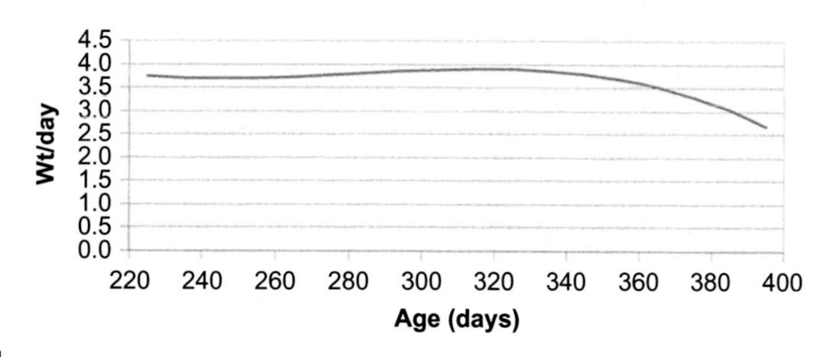

FIG. 6.34

Average daily gains from 230 to 400 days of age.

Courtesy, Dr. Gene Rouse, Iowa State University.

The ultrasound scans for subcutaneous fat indicate that the subcutaneous fat increased as live weight increased. At 600 pounds, the subcutaneous fat thickness was 0.1 in., and at the yearling weight of 1175 pounds, the fat thickness was 0.32 in. (Fig. 6.35). The rate of change in subcutaneous fat thickness (Fig. 6.36) also increased as the bulls approached the yearling weight. This indicates that more energy was used for fat deposition as the bulls approached the yearling weight.

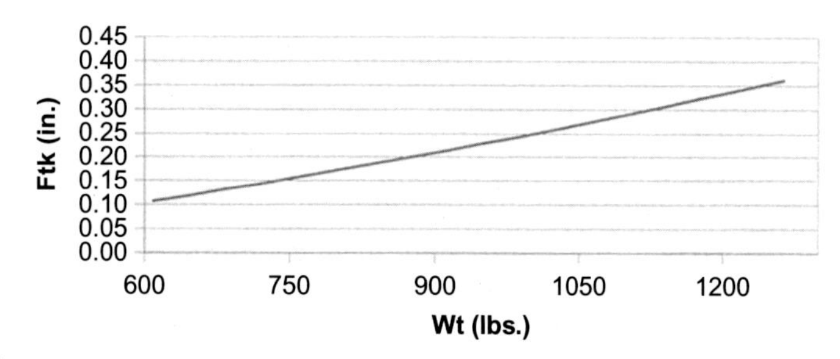

FIG. 6.35

Fat thickness determined by ultrasound scans at the 12th–13th rib for cattle from 600 to 1175 pounds.

Courtesy, Dr. Gene Rouse, Iowa State University.

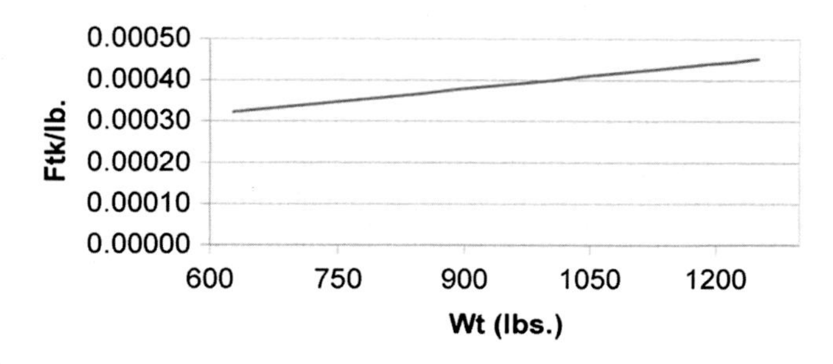

FIG. 6.36

Fat thickness rate of change per pound of live weight gain.

Courtesy, Dr. Gene Rouse, Iowa State University.

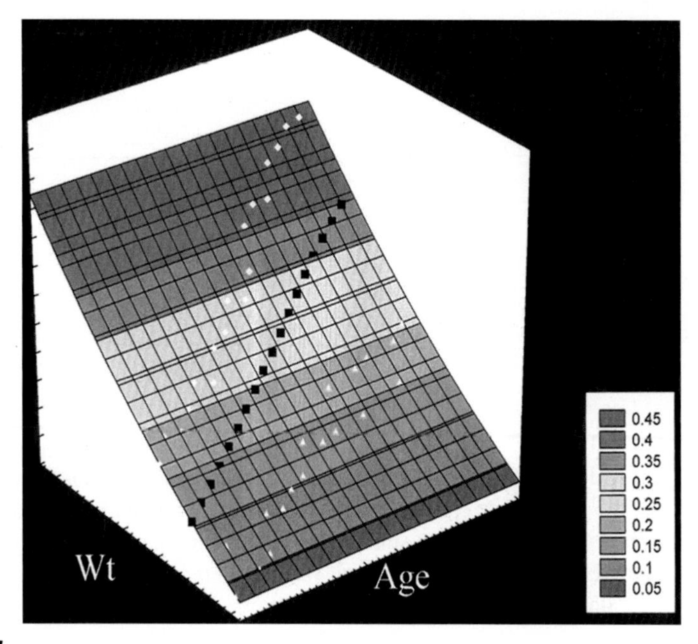

FIG. 6.37

Fat thickness three-dimensional surface plot as a function of age and weight.

Courtesy, Dr. Gene Rouse, Iowa State University.

A surface plot of subcutaneous fat thickness is shown in Fig. 6.37. The surface plot, as a function of age and weight, indicates that weight is the significant factor influencing the subcutaneous fat thickness.

Fig. 6.38 shows the degree of increase in the rib-eye area from 600 to 1175 pounds or at the yearling age. At 600 pounds, the rib-eye area was 7.5 in.^{2}, and at 1175 pounds, the rib-eye area was 12.5 in.^{2}.

The rate of change increase in the rib-eye area, however, declined as the bulls increased in weight (Fig. 6.39). This response, when compared with the increase in rate

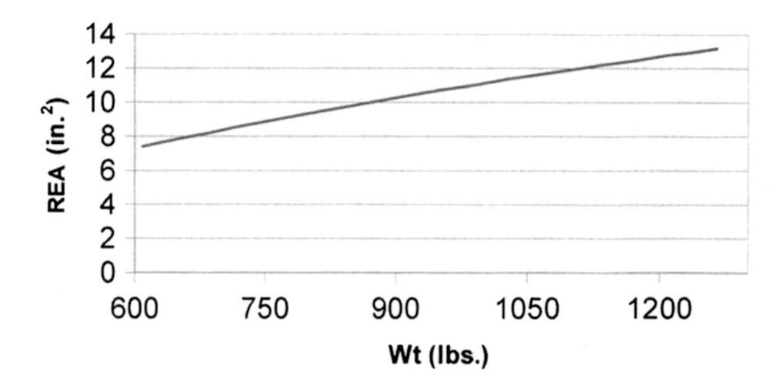

FIG. 6.38

Mean rib-eye area expressed in square inches from 600 to 1175 pounds.

Courtesy, Dr. Gene Rouse, Iowa State University.

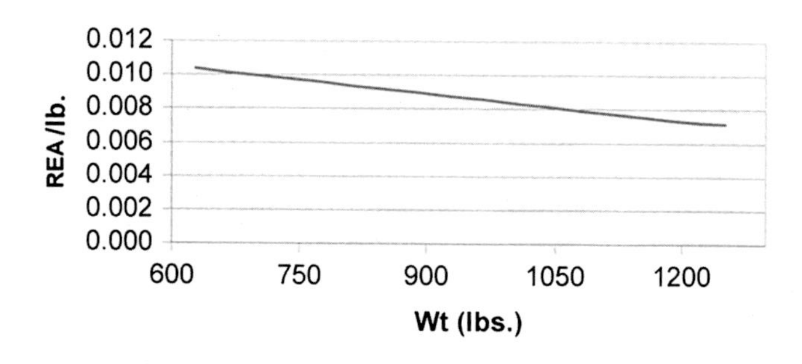

FIG. 6.39

The rate of change in the rib-eye area per pound of live weight gain.

Courtesy, Dr. Gene Rouse, Iowa State University.

for subcutaneous fat deposition (Fig. 6.36), reflects a classic example for the growth and development principle for animal tissue growth. During the growth process, as muscle deposition slows, more nutrients become available for increased fat deposition. This process takes place during the end of the 3rd growth stage and the 4th growth stage.

The three-dimensional surface plot for the rib-eye area (Fig. 6.40) reflects the importance of increased live weight on the increases in the rib-eye area. This is a similar response to the increase in fat cover shown in Fig. 6.37.

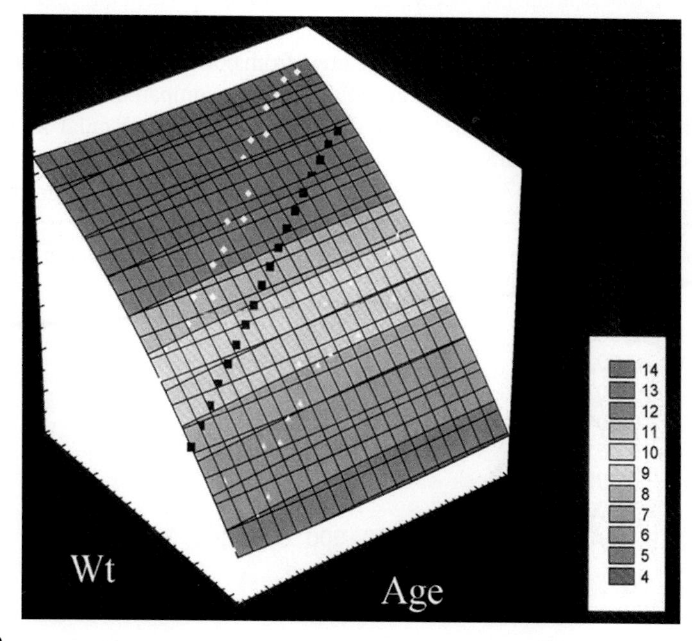

FIG. 6.40

A three-dimensional surface plot for the rib-eye area as a function of age and weight of the cattle.

Courtesy, Dr. Gene Rouse, Iowa State University.

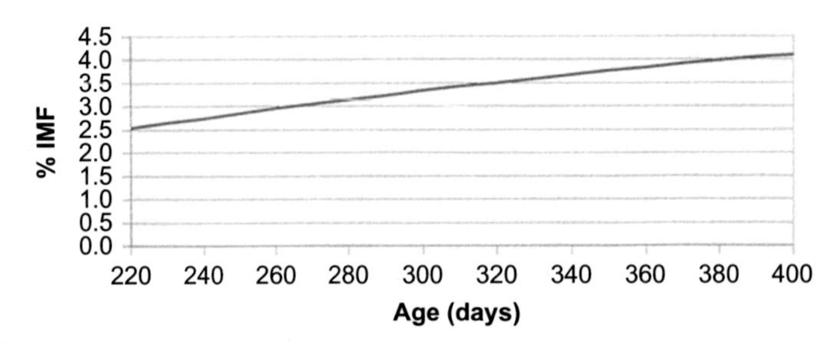

FIG. 6.41

The percent intramuscular fat deposited in the rib-eye muscle from 220 to 365 days on feed.

Courtesy, Dr. Gene Rouse, Iowa State University.

The increases in percent intramuscular fat are shown in Fig. 6.41. The intramuscular fat increased from 2.5% to 3.8% as the bulls grew from 220 to 365 days. During these growth stages, the cattle increased in weight from 600 to 1175 pounds.

The rate of change per pound of live weight gain for percent intramuscular fat was very similar from weaning (220 days and 600 pounds) to the yearling age. At the yearling age, the cattle had average live weights of 1175 pounds (Fig. 6.42). Note that the curve in Fig. 6.42 is nearly flat from 280 to 400 days.

The three-dimensional plot shown in Fig. 6.43 for the percent intramuscular fat as a function of age and weight suggests that the increases in intramuscular fat are more dependent on age than weight as the cattle grew from 600 to 1175 pounds. The difference in deposition patterns for intramuscular fat when compared with subcutaneous fat would suggest that subcutaneous fat (waste fat) and intramuscular fat (taste fat) are independent genetic traits.

The growth curves (6.33–6.43) reflect the association between age and weight for deposition of muscle, subcutaneous fat, and intramuscular fat from weaning weight (600 pounds) to the yearling weight (1175 pounds). Based on the information

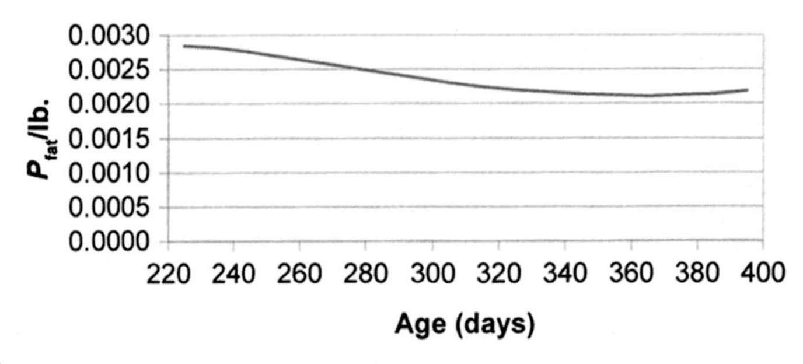

FIG. 6.42

The rate of change for percent intramuscular fat per pound of live weight gain from 220 to 365 days in the feedlot.

Courtesy, Dr. Gene Rouse, Iowa State University.

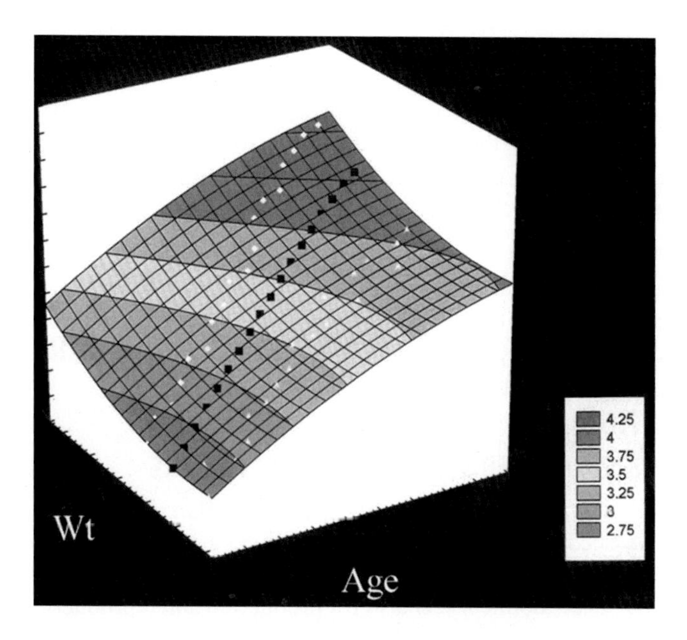

FIG. 6.43

A three-dimensional surface plot for intramuscular fat as a function of age and weight.

Courtesy, Dr. Gene Rouse, Iowa State University.

obtained from the growth curves (Figs. 6.6–6.9 and 6.33–6.43), it is possible to explain the genetic influence on growth of carcass traits in beef cattle.

Muscle tissue matures earlier than fat tissue in the normal growth process. Muscle tissue has nutrient priority over fat tissue when muscle disposition is at maximum growth. When muscle tissue is at maximum growth, cattle at this growth stage, normally achieve maximum growth rate per day of age. This is related to the composition of muscle tissue deposited at this growth stage. Muscle contains a much higher water content than fat tissue and, therefore muscle tissue requires less energy for the deposition of a pound of muscle when compared with a pound of fat.

During the normal growth process for beef cattle, approximately 10 pounds of waste fat (subcutaneous, seam, and internal fat) are deposited for each pound of taste fat (intramuscular). This relationship may partially explain why waste fat reaches maximum deposition after most of the muscle has been deposited.

The intramuscular fat is more dependent on the age of cattle than live weight. Apparently, each day the cattle grow, there is enough energy available to deposit some intramuscular fat. This deposition is regulated by genetic controls for fat deposition and can be very different from one animal to another depending on the genetic traits for intramuscular fat deposition. Because genetic data indicates that subcutaneous fat has separate genetic controls from intramuscular fat and because intramuscular fat is age dependent for rate of deposition, it is possible to genetically select cattle for increased intramuscular fat and less subcutaneous fat.

The data collected from the growth curves (Figs. 6.33–6.43) can be used to develop EPDs for the three independent traits: rib-eye area, subcutaneous fat, and intramuscular fat. The utilization of these EPDs for selection of carcass traits requires all segments of the cattle industry to work together, starting with the Seedstock producer and ending with the packing companies. Management and marketing plans are available that involve cooperation between all segments of the cattle industry.

EXAMPLE OF MANAGEMENT, SELECTION, AND MARKETING PLANS USING GROWTH CURVES AND CARCASS INFORMATION

SEEDSTOCK PRODUCERS AND THE GENETIC BASE

The genetic base for the management and marketing plan starts with the Seedstock producer. The Seedstock producer identifies young sires and replacement heifers with the most desirable EPDs for intramuscular fat, rib-eye area, and subcutaneous fat that fits an economical scenario for production and reproduction traits for a profitable commercial cow herd. Of the three selection criteria, intramuscular fat, rib-eye area, and subcutaneous fat, the subcutaneous fat EPD can be the most difficult to use in a Seedstock genetic selection program because of the relationship between the degree of subcutaneous fat and reproduction traits in female cattle. If the subcutaneous fat becomes too low, the reproduction traits can be reduced. Therefore many Seedstock and cow-calf operators must balance the production traits with carcass traits in their genetic selection programs. For this concern, some cattle producers select for intramuscular fat in their program, and they manage subcutaneous fat deposition in the production portion of their management system.

COMMERCIAL COW-CALF PRODUCTION AND THE GENETIC PLAN

The commercial cow-calf operators produce most of the cattle that enter feedlots in the United States. Therefore these cattle producers have a major responsibility for establishing large numbers of cattle with a low degree of waste fat and a high degree of taste fat. To achieve these goals, the commercial cow-calf producer should use the bulls that are sons of sires with the highest EPDs for intramuscular fat, rib-eye area, and low subcutaneous fat that are within the acceptable frame size for a profitable commercial cow-calf operation. It is also important to use crossbreeding programs that allow traits from different breeds to complement each other for a good genetic base (excellent heterosis). These traits are also important for the replacement heifers generated by the commercial cow herds. These females need to have good genetic traits for growth, muscling, intramuscular fat, and low levels of waste fat so their offspring meet the criteria needed for the marketing plans.

FEEDLOT MANAGEMENT CONCEPTS FOR THE PRODUCTION AND MARKETING PLAN

As cattle in the feedlot approach harvest time, real-time ultrasound measurements can be used to help the feedlot management make good marketing decisions. The three criteria used in the study to establish the genetic base for selection of quality cattle for the feedlots should be used in the decision process on when to send the feedlot cattle to the packer for the harvest process. The three criteria are as follows:

1. Live weight: When cattle reach a weight on the growth curve where average daily gains start to slow, it is a good time to use the real-time ultrasound scans to estimate the carcass traits. This is usually 30 days or about 100 pounds before expected harvest. The live weight of the cattle should be obtained at this time. When the live weight is obtained, the carcass weight of the cattle can be estimated by multiplying the live weight by the estimated dressing percentage (61.5%–63.5%). Marketing cattle at an acceptable weight range is important for obtaining a good market price. Acceptable carcass weights usually range from 600 to 950 pounds. Individual packing companies may deviate slightly from this range.
2. Subcutaneous fat cover: When the live weights of the cattle are obtained, real-time ultrasound measurements for the subcutaneous fat cover should be obtained at the 12th–13th rib area. The fat thickness is a major factor for the determination of the USDA Yield grade. The relationships are presented in Table 6.2. This information can be used by the feedlot management on when to market the cattle to obtain a good market price and not overfeed the cattle. Feeding cattle to a weight where excess waste fat is deposited not only reduces the market value but also reduces the feed efficiency of the cattle in the feedlot.
3. Percent intramuscular fat: The intramuscular fat percentage should also be determined by real-time ultrasound when the cattle are weighed. The intramuscular fat percentage is determined at the 12th–13th rib of the live cattle. Intramuscular fat percentage in the live animal can be related to the degree of marbling in the carcass as cattle grow in the feedlot. An example is shown in Fig. 6.31, Example A. Feedlot management can use this information as a guide on how long to feed cattle to obtain the most profitable return on investments based on the original feeder cattle prices as well as feed costs.

Table 6.2 Relationship between fat thickness at the 12th rib and USDA preliminary yield grade for cattle

Fat cover (in.)	Preliminary Yield grade
0.2	2.5
0.4	3.0
0.6	3.5
0.8	4.0

Marbling score is the primary determinant for the quality (Prime, Choice, Select) grades of beef carcasses, and a major price difference often exists between Select and Choice grades in most markets. Therefore the feedlot managers will balance the value effect of weight, subcutaneous fat thickness (Yield grade Value), and intramuscular fat percentage (Quality grade Value) with feed efficiency of the cattle when decisions are made to market the cattle for harvest to the packer.

The feedlot managers have several options to consider when they market cattle. Marketing cattle on a live weight basis is the old method that has been in place for years. The purchase price (bid) for the cattle may be from a packer buyer or from a price determined at the sale barn. The price is for the live weight and is usually expressed as the dollars per hundred weight. For this option, carcass information is not usually reported back to the owner of the cattle.

Another option for the feedlot managers is to sell the cattle for the value of the hot carcass weight. This is often called selling in the beef by the industry. It is used when dressing percentage is difficult to evaluate, such as muddy conditions of the hide of cattle. Usually this option includes the value based on a truckload of cattle. In this system, the producer or feedlot owner is responsible for the trim loss of the carcass that occurs before the hot carcass weight is obtained. In this marketing option, the packer can obtain the carcass traits such as carcass grade, marbling, and subcutaneous fat and return the information to the feedlot management or owner of the cattle.

A third marketing option is the Grid method. When cattle are sold on the grid option, the packer provides a price for each carcass and provides the 12th–13th rib fat thickness, rib-eye area, and degree of marbling. The carcass Yield grades and Quality grades are also provided.

An example of a Grid Marketing Program that promotes higher carcass prices for cattle that have a Quality grade of Choice and a Yield grade of 1 or 2 will be described later. Only a limited number of cattle have these traits. Therefore a Grid Marketing Program for Choice Quality grades and Yield grades of 1 or 2 can be used by the industry for selection programs to improve the genetic base and strengthen the marketing programs for cattle with a low amount of subcutaneous fat and a high amount of marbling. An example for cattle with these traits is shown in Fig. 6.31 (Example B) when they enter the feedlot until harvest. It is important to market these cattle when they grade Choice and before they deposit more than 0.3 in. of fat at the 12th and 13th rib.

AN EXAMPLE OF GRID MARKETING

This Grid Marketing example is a cooperative program between the JBS Swift Packing Company, the Simmental Association, and the Red Angus Association of America. It is often referred to as the 70 × 70 Grid as it promotes the marketing of cattle from the feedlot when 70% of the cattle grade USDA Choice and 70% of the cattle have a Yield grade of 1 or 2. A Grid used in this type of program is shown in Fig. 6.44.

Plant:	JBS Swift, Grand Island, Nebraska and Greeley, CO
Participation cost:	Scheduling charge = $4/head on all cattle
	Visual tag transfer with QG, YG, HCW = $2/head
Base price:	Nebraska weekly accumulated weighted average week prior to kill
	or forward contract based on CME futures (CMR, Chicago Mercantile Exchange)
TRC base premiums source and age verified	Variable to market conditions, call for current premium: Must be certified through approved PVP program (PVP, processed verified product)
or Not PVP verified	add $.50/cwt dressed to base price
Par price:	Choice YG 3 = base price + TRC premium + clean-up (actual plant performance times grid premiums and discounts. The clean shall never be less than 35% or greater than 75% of the choice/select spread).

Yield grade premium and discounts determied from USDA grades		Quality grade premiums and discounts determined from USDA grades	
	$/cwt	Prime	$6.00
YG1	$4.00	Certified Angus beef	$3.50
YG2	$3.50	Select	USDA box beef cutout with a minium spread of –$3.00 to the Choice
YG3	PAR		
YG 4 variable	(5.00-10.00)*	Standard	($15.00) Back of Select
YG5	($25.00)	Commercial (hardbone)	($30.00) Back of Choice
		Dark cutter	at the market (NWA)
		Others	($30.00) Back of Choice

Carcass weight discounts	
Under 550	($35.00)
1000 up	($15.00)

FIG. 6.44

An example of a 70×70 Marketing Grid for feedlot cattle.

Courtesy, Dr. Gene Rouse.

The management of the JBS Swift Company makes this Grid Marketing Plan available to cattle feeders that purchase feeder calves that are age and source verified. The Simmental and Red Angus Associations work with the commercial cow-calf producers that sell their cattle to the cooperating feedlots. Representatives of the Simmental and Red Angus Associations verify the age and source of the cattle. The Association representatives place tags in the ears of the feeder calves when they are still part of the commercial cow-calf herd. To receive an ear tag from the Breed Associations, the feeder calves had to be sired by a Simmental or Red Angus Bull that has good potential for marbling deposition and a low amount of waste fat in their offspring. The ear tags stay with the feeder calves until they are harvested by the packing company.

THE PACKER RESPONSIBILITIES

The management personnel of the Cooperating Packer for the 70×70 Grid Marketing Program have the greatest responsibility because all of the other cooperating segments require the data collected on the carcasses from the feedlot cattle that were tagged for the 70×70 Grid Program. The Packer collects fat thickness at the 12th–13th rib, intramuscular fat or marbling, USDA Quality grades and Yield grades, and carcass weight. The price paid by the Packer for each individual carcass is based on the USDA Quality and Yield grade as well as carcass weight. The collected carcass information by the Packer Management is sent to the Breed Associations, and the Breed Associations work with the Seedstock, Commercial Cow-Calf Producers, and the feedlot management to distribute all of the carcass and production information collected for the cooperating segments of the industry associated with the 70×70 Grid marketing plan.

The 70×70 Grid Marketing Program provides the information needed to greatly improve the number of cattle marketed for the Choice grade with Yield grades of 1 or 2 by the feedlot management (Fig. 6.31, Example B).

The individual Grid programs are usually developed by the packer to help purchase enough cattle to meet specific markets, such as those requiring a higher degree of marbling for a specific carcass weight range that will fit the marketing criteria for programs such as Angus Certified Beef. The packer can use these Grid programs to pay a premium for carcasses that meet their marketing needs. The premiums paid for the Grid programs provide an incentive for cattle feeders to feed cattle that meet the criteria required to receive the premium price.

The number of Grids available for beef marketing in the United States varies by the individual packer. Packers like to purchase large numbers of cattle for a specific Grid to help with their marketing programs. All of the major beef packing companies have some form of Grids available for the cattle feeder to consider when marketing cattle.

SUMMARY

Growth curve patterns can be used for decisions on animal production practices by farm and ranch managers. The patterns and stages of growth are associated with nutrition and environmental interactions, genetic regulations, and production traits. A common definition of growth is an increase in live weight gain per unit of time and these two concepts are plotted to develop the growth curve.

Growth curves are often divided into four stages. The first stage reflects the early development of the head and neck region as well as the internal organs. The second growth stage results in an increase in body length, major increase in bone and muscle, and organ growth is almost completed. The third growth stage results in a deepening and thickening of the body with a major increase in muscle and the start of significant fat deposition. The fourth growth stage involves greater development of the loin and hindquarter, an increase in depth of the body, an increase in intramuscular fat, the muscle growth greatly declines, and waste fat is the major tissue deposited. Intramuscular fat (marbling), however, is deposited in both the early and later growth stages as an animal grows. Intramuscular fat is more related to daily deposition rather than weight increases as cattle grow to market and mature weights. With the use of ultrasound technology and data from growth curves, excellent management and marketing plans are available for the cattle industry. The management plans include a genetic base for improvement of carcass traits.

QUESTIONS FOR STUDY AND DISCUSSION TOPICS

1. Define postnatal growth for domestic animals.
2. Relate the shape of the growth curve in cattle to production traits.
3. Describe the four growth stages using growth curves.
4. Relate the shape of the growth curve to carcass composition in lambs.
5. What are the muscle, fat, bone, and skin percentages of meat-type and fat-type pigs at 250 pounds?
6. Relate marbling traits in pigs to niche markets for pigs.
7. What is the relationship between deposition of marbling in Angus Cattle and deposition of subcutaneous fat?
8. How can information obtained from growth curves be used to market cattle in feedlots?
9. Describe the Grid concept used to market Choice cattle with a low amount of subcutaneous fat.
10. How can ultrasound technology be used to determine the genetic base for selection of carcass traits in the cattle?

CHAPTER

Harvest processes for meat

7

INTRODUCTION

The process by which livestock are slaughtered to harvest meat has, in many ways changed a great deal in the recent past. In some cases, technological and engineering advances have increased rate of processing, ease of physical labor, and worker safety. Many examples can be made where product improvements have been gained by applying processes to improve quality or safety of the product. Of equal importance are advances in antemortem processes that improve animal handling and animal well-being. The goals of the slaughter process can be summarized as (1) production of safe product, (2) production of high quality product, (3) applications to allow efficient processes, (4) a process that is safe for workers, and (5) ensure good animal welfare during receiving, lairage, and final handling.

SLAUGHTER PROCESS

PREHARVEST HANDLING

Application of proper animal welfare, breeding, and environmental practices are essential for obtaining high quality meat. This begins on the farm where about 50% of the variation in meat quality is due to genetics and management. Furthermore, extreme ambient temperatures have a major negative influence on the physiological behavior of an animal. Rapid fluctuations in the environmental temperature, regardless of the season, are also very stressful to meat animals and ultimately result in inferior meat quality. To be in a high quality condition, meat animals must be well rested and in an appropriate predelivery environment before transport to market. Proper transportation, that is, delivery of meat animals, to the harvest plant is very important. Also, at the market place, there should be compatible holding-pen temperatures, quiet conditions, well-designed and spacious holding pens and walkways, and accessible drinking water for the animals. All of these variables must be controlled before the processes of stunning and exsanguination (blood removal) are carried out.

It is certain that animal genetics can influence meat quality. However, a great proportion of quality variation can be attributed to handling of livestock immediately before slaughter and processing of carcasses immediately after slaughter. It is important to note that workers must know how to properly handle livestock in an abattoir to prevent bruising, blood splash, and ruptured blood vessels leaving blood spots in the meat and death loss.

The Science of Animal Growth and Meat Technology. https://doi.org/10.1016/B978-0-12-815277-5.00007-X

IMMOBILIZATION

In all settings, but especially in commercial operations where space and volume are at a premium, the livestock must be immobilized. This achieves a safer process for the worker and a more welfare-friendly process for the livestock. The approach used for stunning dictates the type of immobilization that is practical. For cattle and lambs, the most common approach is a restrainer conveyor system that is designed to limit the ability of livestock to kick, balk, or turn around. This same type of system is common for swine slaughter when electrical stunning is the method of stunning. Tables 7.1 and 7.2 list practical considerations for determining insensibility in stunned livestock.

STUNNING

Carbon dioxide gas is approved for rendering swine, sheep, and calves unconscious. The gas must be administered in a way that produces surgical anesthesia quickly and calmly, with a minimum of excitement and discomfort to the animals. In the United States, this has become a common approach for swine but not cattle or sheep. Though approved for poultry, no commercial plants are using this approach (as of 2012). The design typically used for swine involves a gondola that will hold six to eight pigs. This gondola is on a lift that drops to a pit containing CO_2. Remember that CO_2 is heavier than atmospheric air. Once the pigs are anesthetized, the gondola rises with the continuation of revolution of the lift. The pigs are removed from the gondola and are ready to be shackled, hoisted, or placed on a table for bleeding.

Table 7.1 Considerations in determining insensibility after stunning

With captive bolt stunned animals, kicking will occur
Ignore the kicking and look at the head. The head must be dead
It is normal to have a spasm for 5–15 s
Eyes should be relaxed and wide open. With captive bolt, there should be no eye movement

Data from the American Meat Institute.

Table 7.2 Signs of a properly stunned animal

Legs may kick, but the head and neck must be loose and floppy like a rag
The tongue should hang out and be limp (a stiff curled tongue is a sign of possible return to sensibility)
If the tongue goes in and out, this is a sign of partial insensibility
When the carcass is hung on the rail, the head should hang straight down, the back must be straight, and it must not have an arched back
A twitching nose (like a rabbit) may be a sign of partial insensibility
When a partially sensible animal is hung on the rail, it will attempt to lift up its head

Data from the American Meat Institute.

The processing facility must maintain a uniform carbon dioxide concentration in the chamber so that the degree of anesthesia in exposed animals will be constant. All gas-producing and control equipment must be maintained in good repair and all indicators, instruments, and measuring devices must be available for inspection by FSIS.

An advantage to this system is that pigs can be handled in groups and not in "a head-to-tail line." Further, physical restraint is not necessary, as the pigs are anesthetized in the gondola. Carbon dioxide stunning is documented to result in better pork quality compared with electrical stunning.

Mechanical stunning is most common for cattle and sheep, especially in commercial plants. There are two types of mechanical captive bolt stunners that may be used to produce immediate unconsciousness in cattle, sheep, goats, or swine. Both types have gun-type mechanisms that fire a bolt or shaft out of a muzzle. The bolt is discharged or propelled by a measured charge of gunpowder (a blank cartridge) or by accurately controlled compressed air. A well-trained and experienced establishment employee must operate both types.

Captive bolts powered by compressed air must have accurate, constantly operating air pressure gauges. The gauges must be easily read and conveniently located for inspection by FSIS. When fired, the bolt in the penetrating type of captive bolt stunner penetrates the skull and enters the brain. Unconsciousness is caused by physical brain damage, sudden changes in intracranial pressure, and concussion. A nonpenetration (concussion) bolt is similar to the penetrating bolt except that it has a bolt with a flattened circular head (mushroom head). When fired, the mushroom head meets the skull but does not penetrate. The animal becomes insensible from the impact or concussion.

The final method approved for stunning animals is electric current. It is approved for use in hogs, calves, sheep, goats, and poultry. It is most commonly used for swine and poultry. Although approved for use in cattle, this is not a common practice. The animal is physically restrained so that the electric current can be applied with a minimum of excitement and discomfort to the animal. Electrode placement can vary. Application of the electrode can be head only, head to chest, or head and back. Head and back application typically results in cardiac arrest. The current passing through the animal must be enough to ensure insensibility throughout the bleeding operation. Blood removal should be as immediate as possible as some animals may regain consciousness. The stun-to-stick interval should be less than 10 s in a commercial plant. If too much current is applied in the stunning process, or if the stun-to-stick interval is too long, petechial hemorrhages or other tissue changes can occur that could interfere with the inspection procedure. These are often referred to as blood splash (Fig. 7.1).

After the animals are stunned, they are shackled and hoisted, usually to an overhead rail, to begin the orderly dressing process (removal of blood, internal organs, head, hide/skin). There is one exception of not stunning animals before removing blood and that is in the religious harvest of beef, such as used in the Kosher process. This is a process carried out according to Jewish law and means proper and clean. The live animal is suspended and the specially trained rabbi, or shochet, cuts across

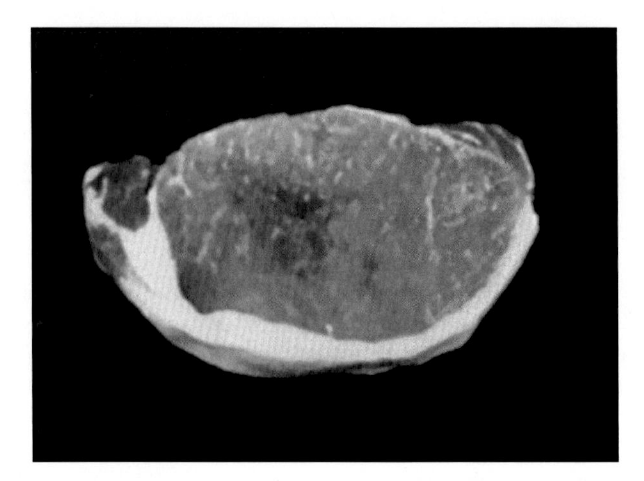

FIG. 7.1

Blood splash in a pork loin chop.

the throat in one continuous stroke with a 14-in. knife. If properly done, the animal becomes unconscious very quickly, in about 3 s.

BRIEF OUTLINE OF CATTLE SLAUGHTER

STUNNING AND EXSANGUINATION

Immobilization is typical in a "V" restrainer on a conveyor. This reduces the ability of the animal to kick, balk, or turn around. Stunning is virtually always a captive bolt. After the animal is shackled and hoisted, exsanguination commences by cutting the hide between the jaw and front of the brisket, followed by insertion of the knife to sever the carotid arteries and jugular vein. Efficient blood removal is a must for sanitation, efficiency, and ultimate product quality.

ELECTRICAL STIMULATION

High-voltage electrical stimulation is often used at this stage of processing to achieve several goals. This process involves passing an electrical current (above 500 V) through the carcass (note that this process is postmortem). The advantages are highlighted in Table 7.3. In general, the high-voltage electrical stimulation, applied at this stage of processing, will improve blood removal and make the hide easier to remove. Other quality advantages from high-voltage electrical stimulation include improved color and tenderness of some muscles.

HIDE REMOVAL/CARCASS WASH

Hide removal is difficult and labor intensive. It is also a critical step in the maintenance of the ultimate safety of the product because the hide is known to harbor harmful pathogens. Careless procedures during the hide removal can result in

Table 7.3 Advantages of electrical stimulation of beef carcasses

Increased blood removal
Improvement in ease of hide removal
Improved tenderness/decreased shear force value
More rapid rigor onset
Brighter color of lean
Shorter required aging period

contamination of the carcass. The primary goals involve removing the hide at the interface of hide and subcutaneous adipose tissue. This process can result in a great deal of contamination of the exterior of the carcass. Briefly explained, the hide removal process involves separation of the hide first at the hind legs, rump, midline (abdominal region) split, and belly skinning. Side pullers are then applied as the very first mechanical aid in the process. Side pullers are hydraulic machines that remove the hide from the sides. Back skinning is achieved manually by separating the hide along the back from a point posterior to the shoulders to a point anterior to the lumbar-sacral junction. At this point, a bar is inserted between the hide and carcass along the back where the hide removal just occurred. This "up-puller" then separates the hide from both hindquarters, leaving the hide only attached at the anterior portion of the carcass. The portion of the hide from the anterior of the carcass is then threaded through a bar called the "down-puller" which, while it turns, removes the hide from the anterior portion of the carcass, including the head. Carcasses are washed and vacuumed throughout this process and at the end of the process with the intent to remove bacteria from the surface of the carcass before their attachment is possible.

SEPARATION OF TRACHEA FROM ESOPHAGUS

This process is completed after exsanguination and is necessary to facilitate evisceration. The esophagus (weasand) must be separated from the trachea and other connective tissue. The separation is called rodding the weasand. After separation, the weasand must be tied or clamped at the cranial end to avoid contamination with the ingesta.

EVISCERATION

Evisceration is completed after head removal, splitting of the sternum, and sealing of the rectum (bung) to avoid contamination of the carcass. Evisceration involves removal of (1) the viscera (rumen, intestines, liver, spleen) and (2) the heart and lungs (often called the pluck).

SPLITTING AND FINAL WASHING AND INTERVENTIONS

Splitting is achieved with a hanging band saw down the center of the vertebral column. Carcasses are generally washed and, in some cases, exposed to steam and or organic acids as interventions to reduce bacterial contamination. Carcasses are chilled for 24–48 h before grading and fabrication.

BRIEF OUTLINE OF SWINE SLAUGHTER

STUNNING, IMMOBILIZATION, EXSANGUINATION

In the case of electrical stunning, swine are immobilized in a "V" restrainer on a conveyor. In this system, pigs are stuck while still horizontal, and only after sticking are carcasses shackled and hung. If CO_2 stunning is used, pigs are moved in groups, stunned in groups, and shackled before exsanguination.

SCALDING/POLISHING/SINGEING

Scalding is the first step in the dehairing option. In commercial operations, the carcasses remain on the rail and are immersed in water. The goal is to denature the protein in the hair follicle to allow for release of the hair. Of course, this requires a combination of heat and time. In general, carcasses remain in the scald tank for 4–7 min at about 140 °F. Dehairing/polishing will remove the hair with the action of rotating polishers. Singeing usually occurs right after the carcasses are rehung through the gambrel tendons. Singeing ensures removal of remaining hair follicles that can be a source of contamination. The temperature of a singeing chamber can be about 1200°C and the surface of the carcass can rise to over 100°C after only a 10- to 15-s exposure.

EVISCERATION

The rectum is tied off with a clamp or band to avoid contamination. The sternum is opened and then the midline is opened. The viscera are comprised of the stomach, small intestine, and large intestine. The pluck includes the larynx, trachea, heart, liver, and spleen (somewhat different than cattle). The viscera and pluck are inspected by an FSIS employee after evisceration, while the carcass of origin can still be identified. Carcasses are split down the center of the vertebral column, washed, and prepared for chilling.

BRIEF OUTLINE OF LAMB SLAUGHTER

IMMOBILIZATION AND EXSANGUINATION

For lambs, the most common approach is a restrainer conveyor system that is designed to limit the ability of livestock to kick, balk, or turn around. Lambs are stunned either with electrical stunning or captive bolt. Blood is removed by drawing the knife through the pelt and severing the carotid arteries and jugular veins. Blood removal can start before or after the carcass is shackled.

PELTING

The hide is removed from lamb carcasses by this procedure. In commercial plants, this takes place while the carcass is suspended by the hind legs. Care is taken to not score the adipose tissue and to avoid contamination from the pelt. In most commercial settings, the hide removal is achieved with knife work around the legs and sternum, and automated systems remove the hide as described with the side-puller, up-pullers, and down-pullers similar to beef hide removal.

EVISCERATION

The trachea and esophagus must be separated to allow abdominal cavity organs and contents to be removed. The esophagus must be banded or clamped to avoid contamination of the carcass with ingesta. The sternum must be split before removal of viscera. As described for other species, the rectum must be tied off to avoid fecal contamination of edible product. Once the viscera and pluck are removed, inspection is conducted by FSIS personnel while the carcass of origin is still identified. Carcasses are washed and rinsed, perhaps also steam pasteurized, and then chilled. Lamb carcasses are not split before chilling.

INSPECTION DURING THE SLAUGHTER PROCESS

It is extremely important to note that livestock, carcasses, and facilities are under constant evaluation for wholesomeness and sanitation during the slaughter process. The USDA Food Safety Inspection Service is responsible for verifying that these conditions are met. Meat inspection systems across the globe and throughout history have a common goal: Ensure safety of meat for public consumption to protect the health of the public. Inspection is generally defined as a third-party verification that the food product is safe for human consumption. Throughout history, science and current knowledge have been applied to ensure this safety. There are many examples of these rules and cultural heritages in antiquity. Though not the focus of this book, important Greek philosophers including Aristotle noted that there was a connection between human and animal diseases. Certainly, we know that inspection was conducted in Europe as early as the 12th century. Most of these early applications of "inspection" were based on observation and health of animals used for food production rather than the product.

In the United States, with centralization of processing and the use of terminal markets in the late 19th century, inspection of livestock was required for products that would be used in interstate commerce. Congress passed a law in 1891 that mandated such an inspection. This inspection (at the federal level) was under the control of the Secretary of Agriculture. It was not until the description of the conditions in a processing plant became known to the public that there was an insistence by the public to require better sanitation. A novel by Upton Sinclair entitled "The Jungle" was written to portray the difficult conditions survived by immigrants in Chicago and other industrialized cities. Sinclair's efforts in the novel depict working-class poverty, the absence of social programs, and the abuse of power by those in charge. Author Jack London called "The Jungle" the "Uncle Tom's Cabin" for wage slavery. Sinclair missed his mark. The public became fixated on the description of the conditions in which their meat was processed. President Roosevelt acted by insisting on an investigation that ultimately led to the 1906 Meat Inspection Act and the Pure Food and Drug Act. The Meat Inspection Act conferred power to the Secretary of Agriculture to make provisions for sanitary handling of meat products and cleanliness of facilities. The Pure Food and Drug Act gave rise to the Bureau of Chemistry which 24 years later was renamed as the Food and Drug Administration.

The USDA Food Safety Inspection Service is headed by the Under Secretary for Food Safety. There are numerous interesting turns and developments of the changes in meat inspection over the years. The primary changes were as follows:

1957 Poultry Products Inspection Act (Provisions similar to 1906 Meat Inspection Act).
1958 Humane Methods of Slaughter Act.
1967 Wholesome Meat Act (Required State Inspection programs to be equivalent to Federal programs).
1968 Wholesome Poultry Products Act (Required State Inspection programs to be equivalent to Federal programs).
1996 Pathogen Reduction Act (Required use of Hazard Analysis Critical Control Point (HACCP), Sanitary Standard Operating Procedures (SSOP), Testing for generic *Escherichia coli*, and establishing performance standards for meat *Salmonella* prevalence).

The federal inspection system is comprehensive and absolute with regard to control of the wholesomeness of meat products. The inspection system includes many obvious requirements, but some in the list might not be immediately evident.

(1) Examination of animals and carcasses for signs of disease or contamination.
(2) Inspection of sanitation of facilities, buildings, grounds, and equipment.
(3) Determining the adequacy of SSOP and HACCP plans.
(4) Assuring compliance with performance standards.
(5) Assuring identification, labeling, and proper disposal of meat that is not fit for human consumption.
(6) Inspection of ALL ingredients used in meat products.
(7) Create and enforce identification standards for inspected meat products.
(8) Require informative labeling to allow transparency to consumer.
(9) Prohibit false labeling (either accidental or fraudulent).
(10) Inspection of imported meat.
(11) Create an educational and awareness program that ensures acceptance of USDA inspected product in global commerce.

To do this, inspectors for USDA conduct inspection in the following areas.

ANTEMORTEM INSPECTION

Inspection begins with antemortem evaluation of livestock on the day of slaughter. Animals that are down, diseased, disabled, or dead are immediately considered unfit for human consumption. Because of concerns for bovine spongiform encephalopathy

(BSE), nonambulatory cattle are automatically eliminated from consideration for slaughter for human food.

Antemortem inspection is conducted by observation of unrestrained animals during lairage. Suspect animals are identified, removed from the group, and isolated for a thorough evaluation. Animals identified as unfit for human consumption are tagged "US Condemned" or killed immediately. Carcasses of condemned livestock must be denatured under the supervision of the inspector. In commercial plants today, the frequency of observations that lead to condemnation is relatively low.

POSTMORTEM INSPECTION

Postmortem inspection is conducted under the supervision of a veterinary medical officer and is conducted to ensure the wholesomeness of the entire carcass and individual products. The lymph nodes are commonly evaluated as a quick indicator of pathogen infection. The physiological role of the lymph nodes is to filter and destroy bacteria. In essence, they isolate infection and prevent the spread of infection. Under normal conditions, these lymph nodes are pale and compact. When the lymph nodes are activated, they become swollen and discolored. If this phenotype is noted, the inspector will note the extent of the presence of disease and determine the condition of the entire carcass. In some cases, removal of affected portions of the carcass is warranted. In extreme cases, the entire carcass must be condemned.

While the carcass is under inspection at several different points in the process, individual stations are available for head inspection and viscera inspection. Inspection of the head and viscera may identify existence of a pathological condition that is detrimental to the wholesomeness of the entire carcass. Because of this, the head and viscera must be inspected while the carcass of origin is still identified. This can be achieved by inspecting the head while it is still attached to the carcass. However, this is not a practical solution for the viscera; therefore the viscera are typically placed on a conveyor that moves in concert with the carcasses on the rail.

Viscera inspection

Inspection of viscera is a valuable approach to identify pathological conditions that detract from the wholesomeness of the product. Lungs, heart, liver, and spleen must be inspected and palpated for such signs of disease and for lesions of previous infection or disease. The lungs are inspected by incision of brachial lymph nodes and mediastinal lymph nodes for signs of infection, tumor, or abscesses. The lungs should also be inspected for obvious lesions caused by infection and pneumonia. Classically, these signs are discoloration, necrotic tissue, vascular congestion, edema, and hemorrhage.

Carcass inspection

Carcasses are inspected for hygiene as well as wholesomeness. Beef carcass inspection includes inguinal, iliac, lumbar, and renal lymph nodes. Carcasses are inspected to ensure that the product has arisen from a healthy animal and that the carcass has not been contaminated during the harvest process. Once a carcass passes inspection,

FIG. 7.2

Example of inspection stamp applied to a carcass during slaughter.

it receives a stamp that states "US Inspected and Passed" (Fig. 7.2). If carcasses are found to NOT be wholesome and therefore unfit for human consumption, the carcass and all associated products are marked "US Inspected and Condemned".

Control of condemned material

Condemned material must not enter the food supply. These products are "tanked" to destroy potentially harmful agents and be used for inedible purposes.

Product inspection

This is a broad area of inspection that considers all components of the product, its ingredients, the sanitation of the procedures, and the HACCP plan.

Laboratory inspection

This area of inspection relates to any analytical assessment that must be used to determine chemical, microbiological, physical, or pathological determinations. This could be done to determine the existence or prevalence of a pathogen or to determine the accuracy of composition.

Marking and labeling

Brands and labels are applied to all products. Inspection legends appear on labels of prepared meat products. Two such inspection legends are shown in Fig. 7.3. Note

FIG. 7.3

Examples of inspection stamps on products.

that they also have an Establishment number. This allows tracking to the processing facility for fresh product. A directory for these establishment numbers is located online (http://www.fsis.usda.gov/wps/portal/fsis/topics/inspection/mpi-directory). A "P" indicates that the establishment processes poultry. A "G" indicates egg product. An "I" indicates an imported product.

Labels must contain the following items: Name of product, ingredients in the product (from most abundant to least), name and address of the processor or distributor, handling instructions, net weight, inspection legend/stamp, and nutritional content panel. In addition, any pictures on the product must accurately depict the product. Finally, there should not be statement to misrepresent the content or origin of the product. In the case of meat products originating from "state inspected" plants, the state inspection legend replaces the requirement of the federal stamp. All other requirements remain the same.

HACCP

In response to the increasing meat-borne illnesses in the early 1990s, the Pathogen Reduction Act was established. This act requires that HACCP be used to ensure safety of each product. This system was designed by the Pillsbury Company in the late 1950s. The first application of HACCP was the development of a food safety system for NASA. HACCP is a system that is proactive, preventative, and science based. It is designed to identify food-borne hazards, determine methods to eliminate those hazards, and systematically monitor processes to ensure elimination of the hazards.

HACCP is designed around seven principles:

(1) *Conduct a hazard analysis.* Hazards may be biological, chemical, or physical. A systematic evaluation of the product and the process can identify the hazards that should be considered. In meat products, there is a significant focus on bacterial pathogens.
(2) *Identify the critical control point.* A critical control point (CCP) is a point in the process where a hazard can be controlled. The key word here is control. This is not simply a surveillance program. A CCP might be a cooking step, high pressure processing, pH reduction, or addition of ingredients to eliminate pathogens (salt, nitrite) or to decrease water activity.
(3) *Establish a critical limit.* This requires that specification for the CCP is set. It could be a specific endpoint temperature, a pH, a specific target water activity, or a particular level of an ingredient like sodium nitrite. Setting a critical limit must be supported by evidence that the critical limit is effective in controlling the identified hazard.
(4) *Monitor the critical control point.* This requires that once the critical limit is established, it must be measured and recorded. An example is measuring and documenting the temperature that was reached during a cooking step. Another might be monitoring the pH if a particular pH was a critical limit.
(5) *Establish corrective actions.* Sometimes this one is called the "what if" step. In other words, "what do we do if the critical limit is not met?" This requires that the decision is already made if the product does not reach the critical limit.

(6) *Establish effective record keeping.* Record keeping is a backbone of a HACCP plan. Records are the documentation that the plan was followed.
(7) *Verify the system.* Verification of the effectiveness of the HACCP plan is required for continuous improvement. This is linked to continuous verification that existing plans and critical limits are controlling the hazards.

INTERVENTIONS IN THE HARVEST PROCESS TO IMPROVE BEEF SAFETY

It stands to reason that good sanitation must be employed during the slaughter process. It has become clear that a thorough understanding of the sources of the pathogens that threaten the safety of the product be applied to develop strategies to reduce their prevalence in meat. This section will focus on applications developed to reduce the incidence of *E. coli* O157:H7 in beef. Similar approaches have been used to reduce pathogens in other species as well.

The source of *E. coli* O157:H7 is most commonly the hide of the cattle. This is a conclusion based on the following: (1) fecal prevalence of *E. coli* is usually less than carcass prevalence, (2) hide prevalence of *E. coli* is in excess of carcass contamination rates, and (3) rates of carcass contamination during processing are highest right after hide removal and then consistently decline during the rest of the process. This information is useful as it provides a strategic target during processing to reduce this pathogen. Preharvest interventions are useful, but evidence suggests that significant contamination or recontamination can occur in the plant livestock holding area. Given this observation, excellent preharvest and production practices targeted at *E. coli* O157:H7 reduction are not effective if hides are contaminated on live cattle upon arrival at the processing facility.

HIDE WASHING

The primary source of *E. coli* O157:H7 is the beef hide. Therefore it makes sense to remove that contamination early in the process of the hide removal. Hides are often washed after exsanguination and before any portion of the hide is opened. Sodium hydroxide (1.5%) is typically used in this wash. Before the dressing process commences, workers will remove any excess liquid or material that was loosened along the hide opening pattern lines. Experiments in commercial settings demonstrate that the hide washing step will reduce the prevalence of *E. coli* O157:H7 on the hides after washing but more importantly on the preevisceration carcass (see Fig. 7.4).

TRIMMING

Operations may decide to trim pattern lines when hide opening occurs to help ensure best dressing procedures. Trimming is also used to remove visible contamination. Trimmers must be properly trained to conduct the trimming in a manner that will

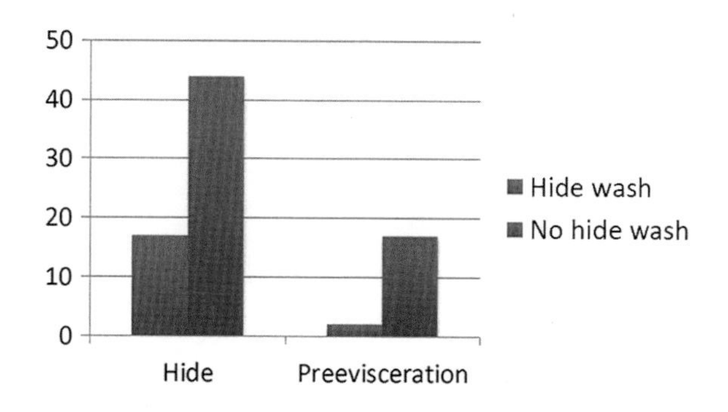

FIG. 7.4

Prevalence of *Escherichia coli* O157:H7 in response to hide washing steps.

create a smooth surface to prevent the formation of flaps or rough surfaces that could decrease the effectiveness of the interventions later in the process. It is also important to ensure proper cleaning and sanitizing of knives and equipment to prevent contamination. Of course, this approach is not useful when contaminated areas are not visible.

STEAM VACUUM

Steam vacuums are a processing aid used to help remove visible contamination. This approach is quite useful to remove contamination along the hide removal pattern line. Hot water (up to 194–197°F) delivered under pressure is effective at loosening contamination and removing bacteria. A common practice today is to apply the steam vacuum process along the hocks and hide removal pattern line.

CARCASS WASHES, ORGANIC ACIDS, HOT WATER, OR STEAM

Whole carcass spray washing is used to remove material contamination during processing. In general, spraying is considered more of a quality process than food safety process in that it prepares the carcass for chilling. However, carcass wash can be useful as a preparation for organic acids, hot water, or steam pasteurization.

Organic acids are useful interventions in the attempt to reduce pathogen loads on the carcass. Lactic acid (4%–5%) is the most common organic acid used. In most cases, it is sprayed on hot (approximately 130°F). Lactic acid is considered a processing aid and, therefore products produced from carcasses sprayed with lactic acid do not need to include lactic acid on the ingredient list. Hot organic acid washes can result in a 1–2 log reduction in bacterial load.

HEATING STEPS (HOT WATER OR STEAM PASTEURIZATION)

Not surprisingly, hot water (165°F) is a very useful intervention to reduce pathogen loads. Maintenance of this temperature is necessary to ensure that effectiveness is maintained. Research has demonstrated that steam pasteurization in a commercial plant results in a 1–2 log reduction in aerobic plate counts.

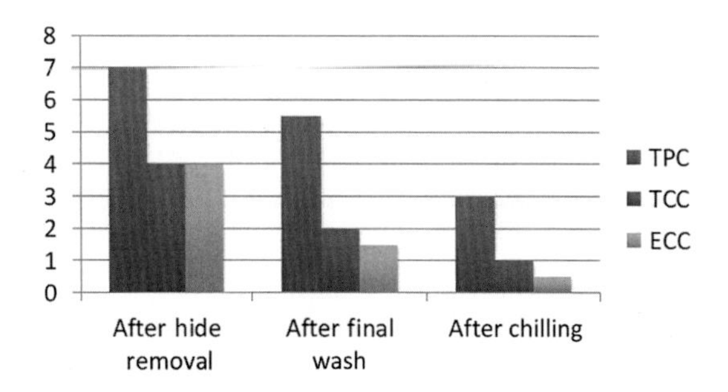

FIG. 7.5

Log value of total plate count (TPC), total coliform counts (TCC), and *E. coli* counts (ECC) on beef carcasses during beef processing.

None of the aforementioned approaches are 100% effective. The commonly accepted approach today is to use multiple sequential or layered interventions to reduce the risk of contamination of pathogens. Each firm and even each processing plant has unique approaches to achieve this reduced risk, but virtually all employ a multiple hurdle approach. A multiple hurdle approach takes advantage of additive effects of interventions to improve food safety. Fig. 7.5 demonstrates that when interventions are employed throughout the slaughter process, total plate counts, total coliforms, and total *E. coli* counts are all reduced.

SUMMARY

The primary objective of the slaughter process has not changed for centuries. However, new requirements to ensure product safety and quality while ensuring animal welfare have resulted in innovation and technology. It is certain that improvements in all of these categories can still be made with integration of expertise by meat scientists, engineers, microbiologists, and animal welfare experts.

QUESTIONS FOR STUDY AND DISCUSSION TOPICS

1. Define three stunning methods.
2. How does CO_2 stunning affect how livestock are handled before exsanguination?
3. Why is the esophagus clipped before evisceration?
4. How long and at what temperature are pigs scalded to remove hair?
5. What is meant by a “multiple hurdle approach” for food safety?
6. Define the Following Terms in the context of animal slaughter: Immobilization, Evisceration, Pluck, Weasand, Electrical stimulation, Steam vacuum, Carcass washing, Multiple hurdle interventions for pathogen reduction

CHAPTER

Methods to measure body composition of domestic animals

8

INTRODUCTION

Considerable variation exists in body composition of all domestic animals. The variation depends on several factors, including the stage of growth, nutritional history, genetic background, and environmental conditions. The body composition of domestic animals determines the economic value of animals produced for meat production. Individuals working in the animal industry need accurate and efficient methods to estimate body composition when animals are sold for slaughter, selected for breeding animals, and selected for placement in feedlots. People working in the livestock and meat industry have several choices to estimate body composition. These choices will be presented in this chapter.

SUBJECTIVE EVALUATION OF ANIMALS

VISUAL EVALUATION

Visual evaluation of live animals is the historical method of choice for selection of superior animals. For more than 100 years, good objective methods were not available, and livestock judges and livestock buyers were trained to become experts for selection of breeding animals and live-animal evaluation for animals sent to market. Some animal scientists became world leaders in the livestock industry because of their ability to estimate body conformation and carcass traits in domestic animals.

Domestic animals have major differences in conformation, and these differences are associated with the percentage muscle and fat in the carcass. Examples of the conformation differences in the pig are shown in Fig. 8.1. Observing pigs from the back view reflects major differences in conformation when muscular and fatter pigs are compared. When evaluating animals for carcass traits, the first observation should be at the ham region of the pig, the leg region of the lamb, or the round region of the beef animal. Animals that have more muscling will be wider through the center of the ham, leg, or round when compared with fatter animals.

Muscular animals will also have more width over the top of the rump section of the body. Very muscular animals will have a significant contour shape over the loin region, and fatter animals will have a more flat and angular appearance over the top of the loin. Fig. 8.1 reflects these traits in the pig. These conformation characteristics

The Science of Animal Growth and Meat Technology. https://doi.org/10.1016/B978-0-12-815277-5.00008-1

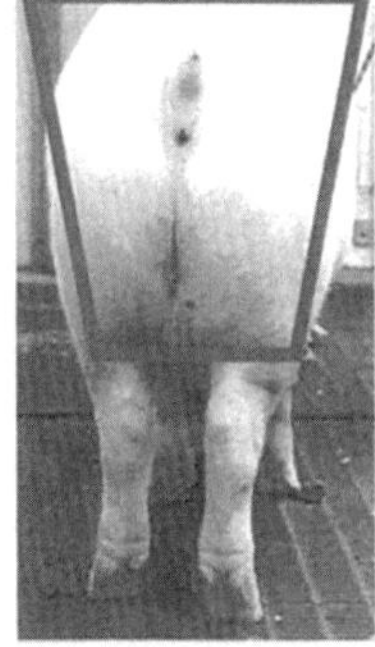

FIG. 8.1

This example reflects differences in body shape of pigs with good and poor muscling.

Courtesy of the National Pork Board, Des Moines, Iowa.

in animals represent desirable and undesirable traits and are often observed by individuals who use visual methods to predict carcass traits and carcass value.

Cross section illustrations of the ham region of pigs (Fig. 8.2) illustrate the muscle traits for the muscular and fatter pigs shown in Fig. 8.1. Compare the shape of the good muscling pig that is lean with that of the poor muscling pig that is

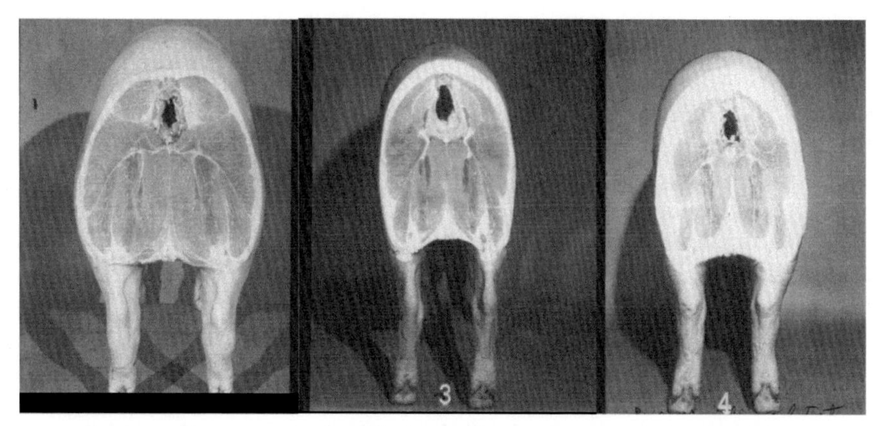

FIG. 8.2

This figure illustrates the relationship of the shape of the ham and loin region from the back view of the pig to the degree of muscling and fat in the pork carcass.

Courtesy of the Animal Science Department, Iowa State University.

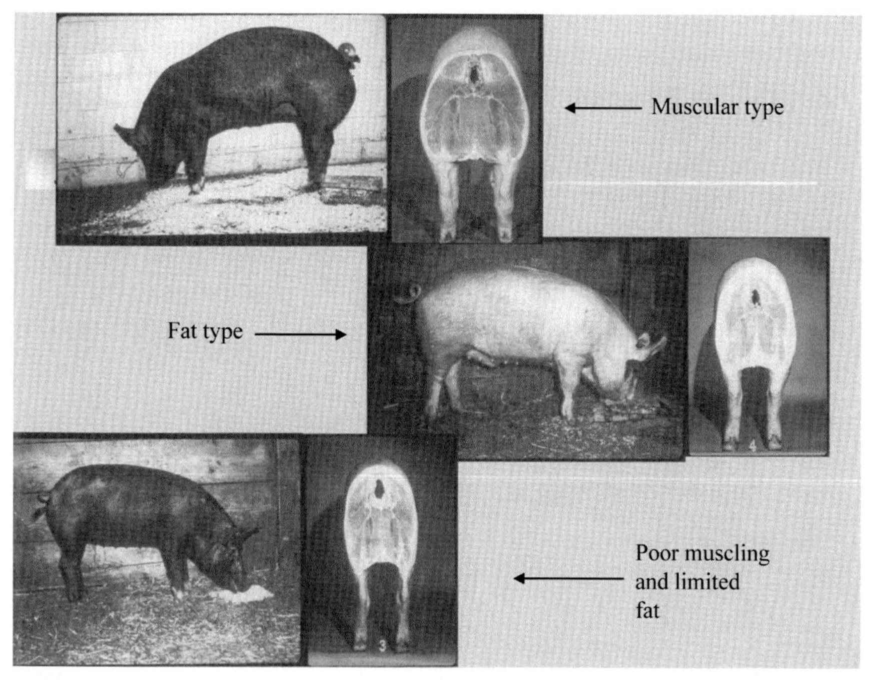

FIG. 8.3

This figure illustrates the relationships of conformation differences in the pig with conformation differences in the cross section of the carcass.

Courtesy of the Animal Science Department, Iowa State University.

also lean. Fig. 8.3 relates the shape of the live animal with the shape of the carcass cross sections indicating major conformation differences in width and depth of the ham and more contours over the loin for muscular pigs and more flat conformation patterns for the fatter animals. Fig. 8.4 indicates similar contrasts for cattle with two levels of fat thickness. In general terms, the conformation concepts used for live-animal evaluation of muscling and fat are very similar for cattle, pigs, and sheep.

Individuals who are highly trained for live-animal evaluations such as judges for livestock shows and livestock buyers can do a good job in the evaluation of conformation and fat thickness. For muscling and fat thickness, they have an accuracy of 70%–75%. Objective methods such as ultrasound have a higher accuracy (90%) to estimate fat percentages and muscling. Therefore the ultrasound technology is used extensively, and it is a more accurate technique to estimate carcass meat and fat percentages. If objective methods such as ultrasound are not available or are not practical to use, visual and subjective evaluations combined with live weight of the animal can provide good predictions of body composition by well-trained evaluators.

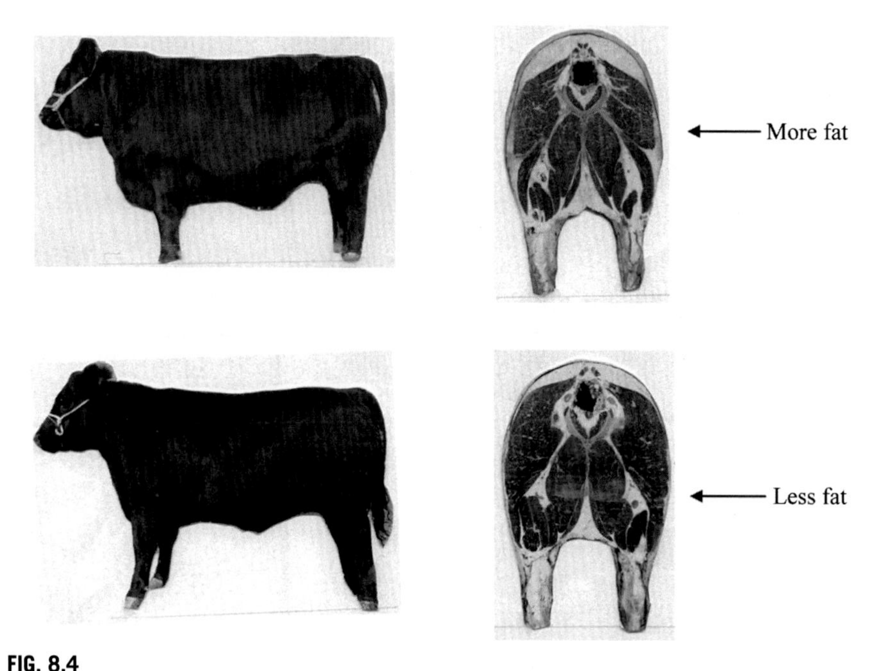

FIG. 8.4

Example of cattle with two types of conformation and fat thickness.

Courtesy of the Animal Science Department, Iowa State University.

VISUAL EVALUATION METHODS FOR MARKETING BEEF CATTLE

Some beef cattle are still purchased by cattle buyers who represent the meat-packing companies. The cattle are bought on the basis of live weight and visual evaluation of muscling and fat percentages in the carcass as well as the expected or estimated marbling in the carcass. Marbling is often associated with the breed of cattle, and the Angus breed is recognized to have greater marbling levels than are other beef breeds. Marbling variation can also be great within breeds as demonstrated by the variation in marbling expected progeny differences (EPD) for Angus sires reported in the sire summary database. Without knowledge of the genetic potential for marbling in beef cattle, it is difficult for the beef-cattle buyer to estimate marbling in live animals. Feedlot cattle that are crossbreds can be even more difficult for cattle buyers and feedlot managers to estimate the quality grade and live value of the cattle. Therefore options are available and cattle can be purchased by the packing company using three options depending on the interest of the packer and the cattle feeder.

LIVE WEIGHT AND VISUAL EVALUATIONS

This marketing option occurs when a packer buyer or sale-barn manager provides a bid on a load of live cattle. The bid is based on a price per cwt (hundred weight) when the quality and yield grade are estimated as well as the dressing percentage. This is a traditional method, but it also can be a higher risk method for both the packer and the cattle feeder if the visual estimates for quality grade, yield grade, and dressing percentage are not correct. An example would be cattle for the Certified Angus Beef Program. Since the development of the branded beef programs, Certified Angus Beef (CAB) became the most successful of the branded beef marketing programs based on a higher degree of marbling for the CAB brand. When the cattle are sold from the feedlots, the premium paid for the CAB brand is often based on the black color of the hide. Because premiums are paid for the black cattle due to the popularity of the CAB brand, several cattle breeds developed the black hair color to receive a premium when the cattle are sold on a live weight basis; therefore many black crossbred cattle may not be predominantly Angus. This increases the difficulty of live-animal evaluation for marbling traits.

OTHER MARKETING OPTIONS

PRICE BASED ON THE HOT CARCASS WEIGHT

When selling cattle on the basis of the hot carcass weight, some cattle buyers will refer to this method as "in the beef" selling. This marketing option removes the risk for estimating dressing percentage by the packer buyer. A practical example for this option would be the elimination of the packer buyer to estimate the "mud-weight" on the hide of the cattle, which often occurs in the Midwest states during the wet seasons. Also in this system, the cattle feeder or producer is responsible for the trim loss that occurs before the hot carcass reaches the scales to obtain the carcass weight. An example may be the trim loss to the carcass during inspection if the USDA inspector removes a portion of the carcass before the carcass weight is obtained. In this method of selling cattle, the cattle feeder also stands the risk of condemnation of the entire carcass by the USDA inspector if disease is detected in the carcass during the inspection process. The packer provides the cattle feeder a price based on the value of the hot carcass weight for the load of cattle marketed.

PRICE BASED ON GRID MARKETING

The grid marketing concept is based on the individual carcass price paid by the packer using hot carcass weight, quality grade, and yield grade. Premiums or discounts for each individual carcass are determined based on the optimal value for the carcass quality grade, yield grade, and carcass weight. An example of a marketing grid option is presented in Table 8.1. Note the price adjustments are from the base price near the center of the grid (Low Choice and Yield grade 3).

Grid marketing is a method to price individual beef carcasses based on the carcass merit. The examples for grid marketing are from Dr. Gene Rouse, ISU. The base price or "par" is the price/cwt carcass that would be paid for 100% Choice, Yield grade 3 carcasses. Premiums and discounts can occur when yield grades are above 3 or below 3. Yield grades above 3 are discounted, and yield grades below 3 are given

Table 8.1 An example of a grid for marketing beef cattle

YG[a]	Select	Choice –	Choice +	Prime	Carcass weight discounts[b]	
1	(4.00)/126.00[c]	4.00/134.00[c]	8.00/138.00[c]	14.00/144.00[c]	Under 500 lb	(40.00)
2	(5.00)/125.00	3.00/133.00	7.00/137.00	13.00/143.00	500–549 1b	(15.00)
3	(8.00)/122.00	Base 130.00	4.00/134.00	10.00/140.00	950–999 1b	(8.00)
4	(28.00)/102.00	(20.00)/110.00	(16.00)/124.00	(10.00)/120.00	1000+ lb	(20.00)

[a]*Yield grade.*
[b]*Examples of carcass weight discounts. These discounts are subtracted from the grid carcass price. For most packing companies, the acceptable carcass weight range is 600–950 pounds. Carcass weights outside of this range will be discounted.*
[c]*Carcass value example for each quality and yield grade adjusted from the base value of 130.*

Table 8.2 Pricing cattle using the Grid concept

Item	Value
Example one	
Quality grade of the carcass	Choice
Yield grade of the carcass	2
Carcass weight (pounds)	950
Carcass value obtained from the Grid	
Choice, 2	$133.00/cwt carcass
Carcass weight adjustment	$8.00/cwt carcass
Carcass value	$125.00/cwt carcass
Example two	
Quality grade of the carcass	Select
Yield grade of the carcass	1
Carcass weight (pounds)	542
Carcass value obtained from the Grid	
Select, 1	$126.00/cwt carcass
Carcass weight adjustment	$15.00/cwt carcass
Carcass value	$111.00/cwt carcass

a premium. Quality grades below Choice, such as Select, are given a discount, and quality grades above low Choice can be given a premium. Carcass weight is also important in grid marketing. The acceptable carcass weight for most packers ranges from 600 to 950 pounds. Some packing plants may vary from this range. The carcass value can be reduced if the carcass weight is below or above the range listed in the grid. See Table 8.1. The discount is determined by the pricing policy of the individual packer. Examples for pricing cattle using the Grid concept for marketing from the feedlot to the packing company are presented in Table 8.2.

APPLICATION OF GRID MARKETING TO FEEDLOTS

Fortunately, objective data can be used to help in the decisions by feedlot managers for Grid marketing of cattle. Ultrasound technology is now available to estimate fat thickness and marbling in beef cattle before they are harvested. Therefore this technology can be used in marketing decisions for the grid option.

The livestock manager of feedlots should monitor on a regular basis the rate of gain and performance traits of the animals on feed and relate this information to growth curve data (Chapter 6) to determine efficiency of live weight gain and body composition of the animals in the feedlot. The live weight of the cattle, dressing percentage, carcass fat thickness, and marbling can be influenced by several factors such as the frame size, ration energy level, animal age, temperature, humidity, and even subclinical diseases. Therefore livestock feeders should take these factors into consideration and use standard weighing and management procedures when live weights are obtained to determine the time to market cattle. Currently, there are feedlot monitoring programs that can project the weight of cattle with increased time on feed taking type of cattle, energy level, and weather into consideration.

VISUAL EVALUATION FOR CARCASS TRAITS

USDA grading for beef carcasses

The USDA Beef Grading System uses visual and some objective evaluation methods to determine the carcass quality and cutability grades. The USDA grades for beef carcasses are established by trained USDA graders. They can combine visual and more objective methods such as photo imaging to determine the degree of marbling, fat thickness, kidney fat, and carcass maturity to determine the USDA grade that reflects the carcass value for Prime, Choice, and Select grades.

USDA pork carcass grading

The USDA pork carcass grading system has never been used by the major packing companies in the United States. The pork slaughter companies developed their own grading system based on linear measurements for backfat thickness, degree of muscling, and carcass weight, and the packing company pays the pork producer the values of the carcass based on their own grading systems. The major pork processing companies currently use high tech, high speed, and excellent objective methods to determine the carcass value (usually the percentage muscle in the carcass) for payment to the pork producer.

USDA lamb grading

The USDA lamb grades are based on the quality traits and yield of retail cuts. There are two separate grades: Yield grades (1–5) to predict retail cuts and quality grades (Prime, Choice, Good, Utility). These grades are determined mostly by visual subjective methods used by trained USDA graders.

The linear measurement for fat thickness estimated by visual observation or measured over the 12th–13th rib area (Fig. 8.5) is used to determine the Yield grade. The carcass quality grade is determined by visual evaluations for bone maturity, color of the flank muscle, fat streaking in the flank muscle, and conformation of the leg (Fig. 8.5). Therefore lamb values are based on subjective methods rather than objective methods that are used in the pricing of pork carcasses.

Evaluation methods for meat judging contests

Visual evaluations are the major methods used for Collegiate Meat Judging and Carcass Evaluation contests in the United States. The contests are an excellent approach to understand fundamental concepts for carcass evaluation and grading. The contests are also enjoyable for the students who compete at the collegiate level. The contests provide self-confidence and leadership skills and are rewarding when an individual or team wins a division of the contest or the overall contest.

Carcasses in Fig. 8.6 reflect conformation differences in pork carcasses. The five examples range from excellent muscling to poor muscling. Examples of this type are often used for visual training of individuals to evaluate carcasses for conformation

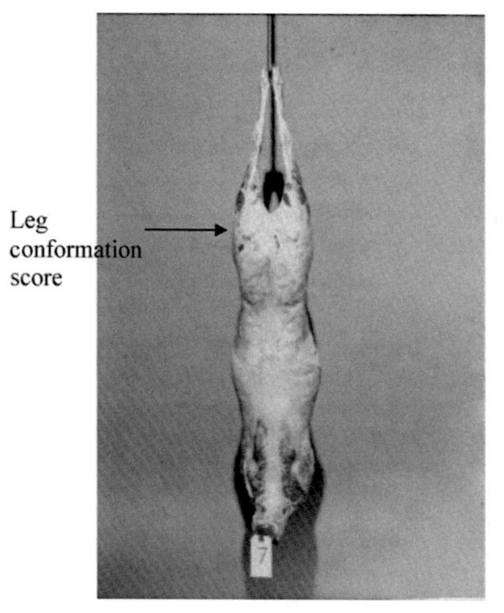

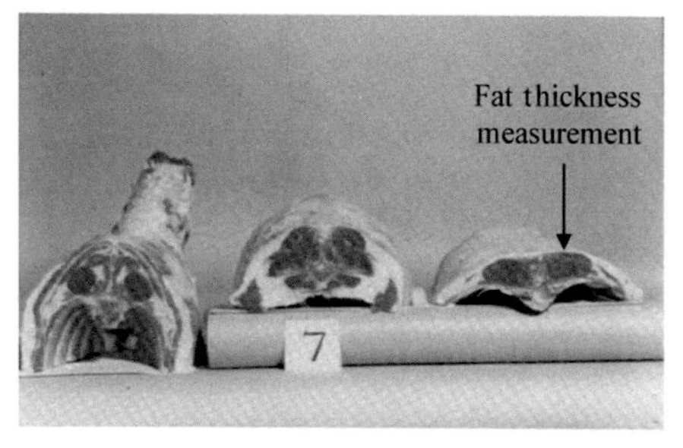

FIG. 8.5

An example of linear measurements for fat thickness and conformation score for USDA lamb carcass grades.

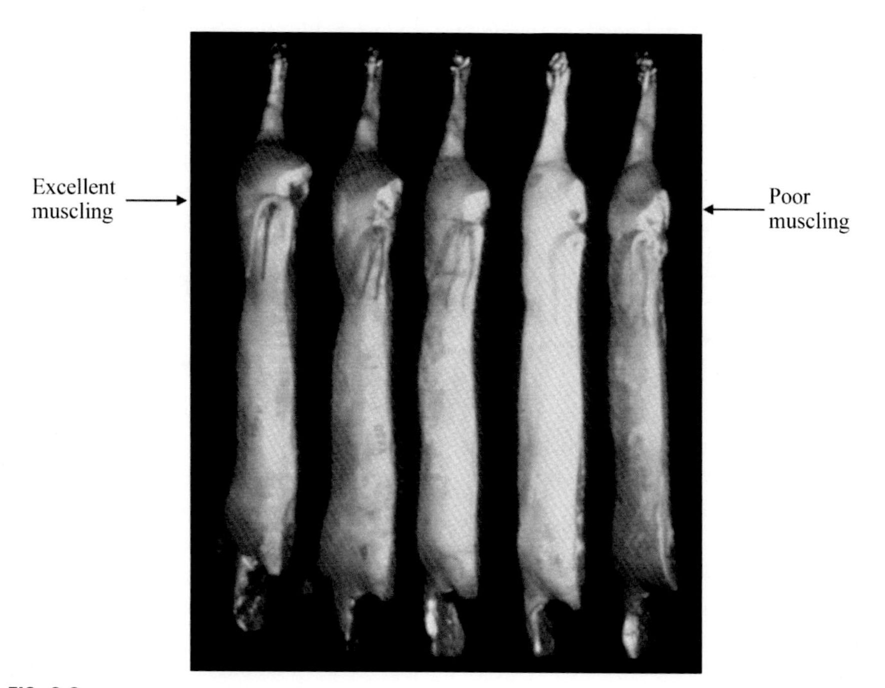

FIG. 8.6

Example of five pork carcasses with different conformation from excellent to good to poor amount of muscling.

Courtesy of the USDA.

and muscling traits when preparing for evaluations contests. The relationship between carcass conformation scores and retail value of the carcass is not as high as the fat thickness measurements and retail value, but it is still an important component of the visual evaluation methods.

Fat thickness in the carcass is often estimated by visual observations in judging contests, even though objective methods are more accurate than the visual or subjective methods. Under certain conditions, however, objective methods may not be available, and then visual evaluations are used. Under these conditions, the experience from judging contests can be useful in estimating carcass value.

LINEAR MEASUREMENTS

LIVE-ANIMAL MEASUREMENTS

Linear measurements used for evaluation of live animals were the only methods available in the early 1900s to evaluate animals for body composition traits. Some objective methods, however, were developed from 1900 to 1950, but they were not practical methods. Therefore livestock continued to be evaluated by simple linear measurements such as length, width, height, and circumference of the live animals and carcasses. Reference points were related to specific anatomically defined locations on the skeleton. Length of the animal was probably the most used linear measurement. Fig. 8.7 shows the length measurement used for pork carcasses. Relating the carcass length measurement to the estimation of length in the pig is shown in Fig. 8.8. Many people who evaluate pigs for body length will compare pigs from the center of the ham to the center of the shoulder. The relationship between carcass length and body composition for carcass meat percentage is low, but the linear measurements for length of live animals can provide some value and insight into the variations of frame and skeletal size of animals and the relationship of skeletal size to mature weight, growth rate, and breed characteristics in genetic selection programs.

BACKFAT PROBE

This section provides a short history about the backfat probe as it provided a significant and simple method to reduce fat in the genetic selection programs for pigs and cattle. It was developed by Dr. Lanoy Hazel, Iowa State University, in 1952. It was the method of choice to measure fat thickness in the pig for more than 40 years. After 40 years, good ultrasound technology was developed and provided a more accurate process to measure both backfat thickness and loin eye area by obtaining just one measurement with the ultrasound machine. The backfat probe, however, was the tool responsible for developing meat-type pigs in the United States. Therefore the probe has much historical significance. If a pork producer could not afford the high costs of an ultrasound machine, the backfat probe can still be used for a genetic selection program to reduce fat in breeding animals. Therefore a detailed description for probing pigs is provided in this chapter.

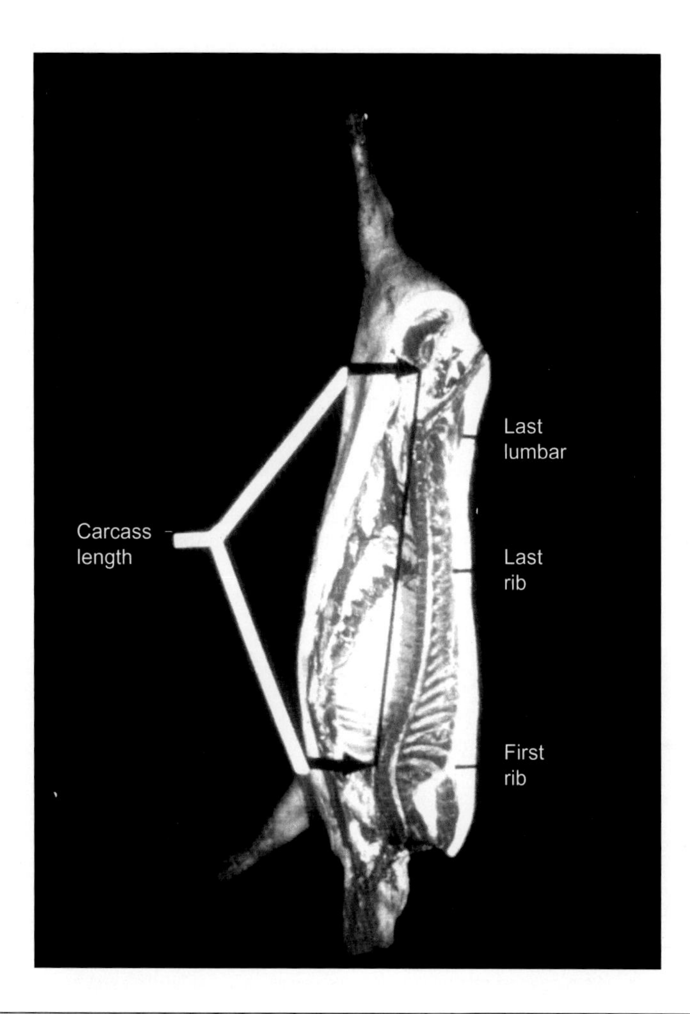

FIG. 8.7

An example of linear measurement locations in the pork carcass to measure carcass length and backfat thickness. Carcass length is measured from the first rib to the aitchbone in the ham. Carcass backfat is measured at the last lumbar, last rib, and first rib.

Courtesy of the USDA.

Probing a pig for backfat thickness

When probing a market-size pig for backfat thickness, the probe is inserted into the subcutaneous fat layers by making a small incision (Fig. 8.9) with a scalpel (Example B) approximately 2 in. off the midline of the back of the pig at three locations (Example A) as shown in Fig. 8.9. These three locations give the best predictor of subcutaneous fat thickness because these locations in the loin regions also reflect the amount of

FIG. 8.8

An example of the location to estimate length in the pig, from the center of the ham to the center of the shoulder.

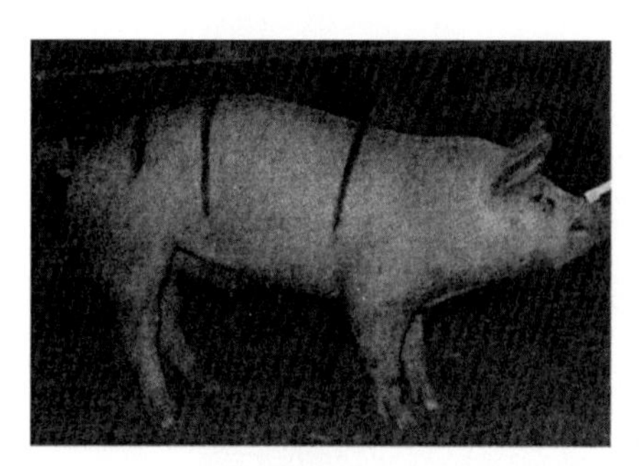

EXAMPLE A
The 3 historical points of a hog on which backfat measurements were made, 2 in. off the midline—the 1st rib, last rib, and last lumbar vertebra.

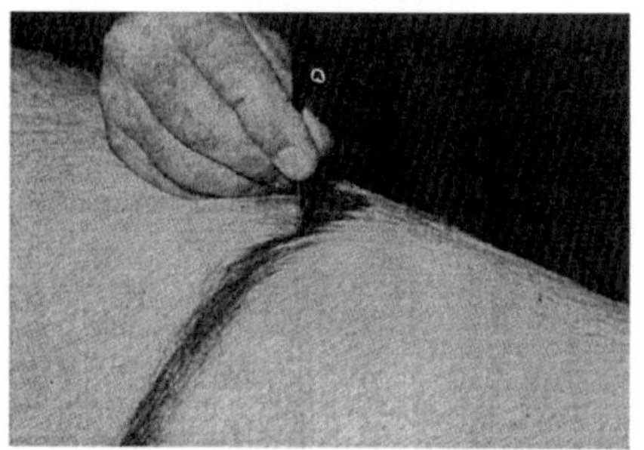

EXAMPLE B
A metal ruler is inserted through a cut in the skin, perpendicular to the length of the pig and about 2 in. off of the midline down the loin muscle. At that point the depth is measured.

FIG. 8.9

Illustration for probing a market-size pig.

Courtesy of Maynard Hogberg, Iowa State University, Ames, Iowa.

muscle in the loin of the pig. An example is shown in Fig. 8.10. Note the shape of the small loin eye muscle and backfat depth in a pork carcass representative of pigs in the 1950s when the backfat probe was developed. This reflects the genetic base of pigs for fat deposition in the 1950s. The progress made in reducing the fat deposition in the pig by using the backfat probe is illustrated in Fig. 8.11. Animal C in Fig. 8.11 represents the 1950-type pig and animal A represents a muscular pig with low backfat and excellent muscling resulting from good genetic selection methods using the backfat probe.

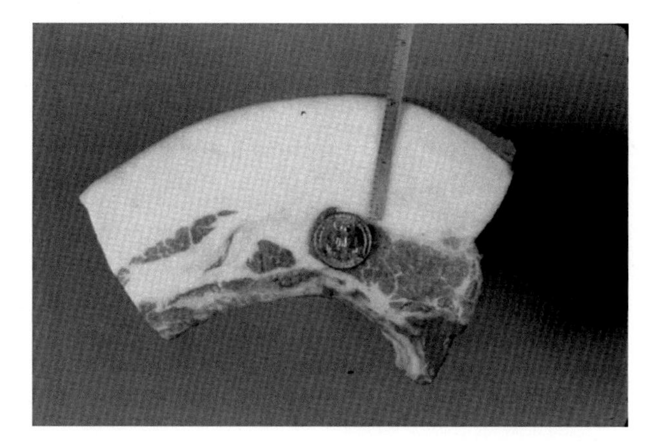

FIG. 8.10

An example for the location of the backfat probe to measure fat thickness in a pig that is representative of pigs in the 1950s when the backfat probe was developed.

Relationship of loin-area shape and backfat thickness, which is important when probing a pig for backfat thickness

A large longissimus muscle in a pig with a small amount of backfat results in a "butterfly" shape to the loin when observed from the back of the pig (Fig. 8.12). Very muscular and trim pigs have a noticeable groove down the center of the back when viewed from the back of the pig. The groove develops from a very small amount of fat in the loin region and a very large loin eye muscle resulting in a butterfly shape to the loin region. Pigs with more fat and a smaller loin eye area have a flatter and angular appearance to the loin region when viewed from the back of the pig. Examples are shown in Fig. 8.11, animal B. Fig. 8.13 shows the location of the backfat probe when it is inserted in the carcass of a muscular and trim pig that had a butterfly shape to the loin region. Compare this example with the fat-type example in Fig. 8.10.

Backfat probe procedures to use when inserting the probe

When the backfat probe is used to measure the subcutaneous fat depth on a pig, it is important to remember that the subcutaneous fat has three individual layers of connective tissue (Fig. 8.14) between the backfat layers. The backfat probe must be inserted through the first two layers of connective tissue before the third layer is reached. The third layer is located over the longissimus muscle. Once the incision through the skin is made, a small and narrow metal ruler or needle is placed through the "false lean" or aponeurosis connective tissue separating the outer and middle layers of subcutaneous fat. The probe should continue to be inserted through the fat layers until it reaches the epimysial connective tissue that covers the longissimus muscle. This is a very thick connective tissue compared with the aponeurosis layer. Once the probe reaches the epimysial connective tissue layer, the depth should be visually verified and recorded.

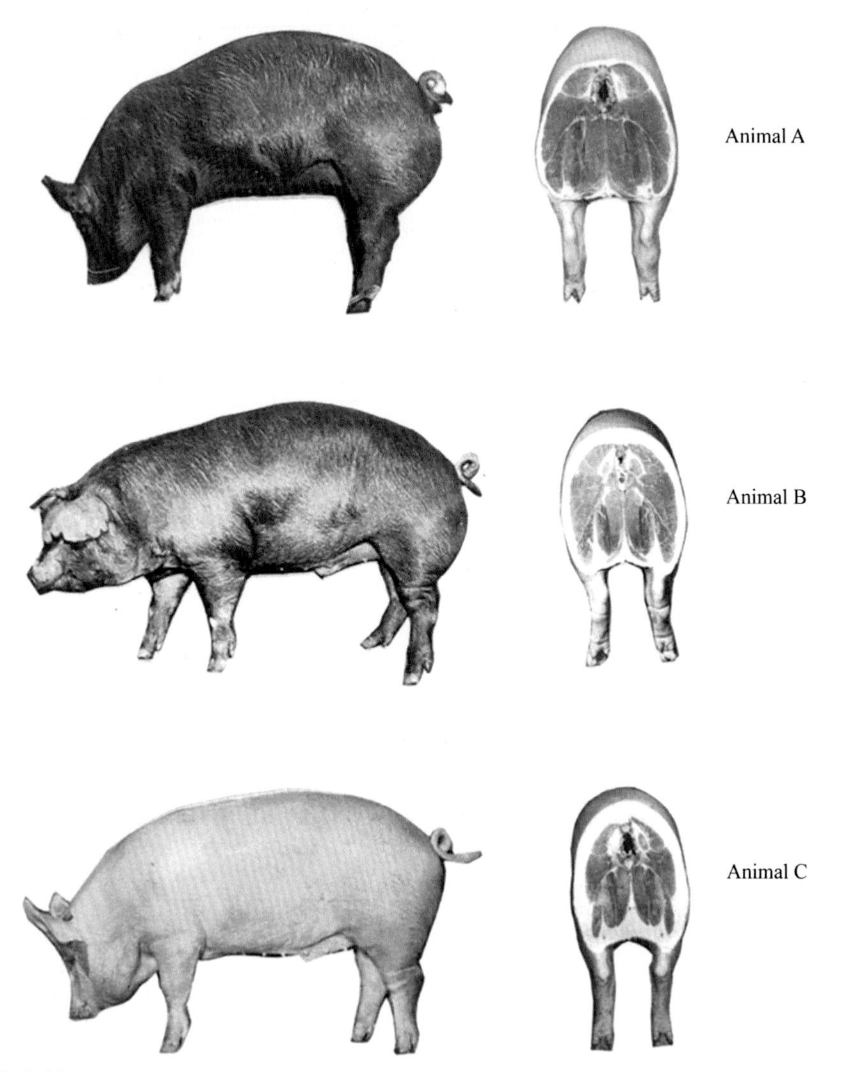

FIG. 8.11

Example of fat and muscling traits in pigs when the backfat probe was used to reduce fat in the pig population. Animal C represents the 1950-type pig, and animal A represents the genetic base for lines of pigs developed by the use of the backfat probe to establish the meat-type pig.

Courtesy of the Animal Science Department, Iowa State University.

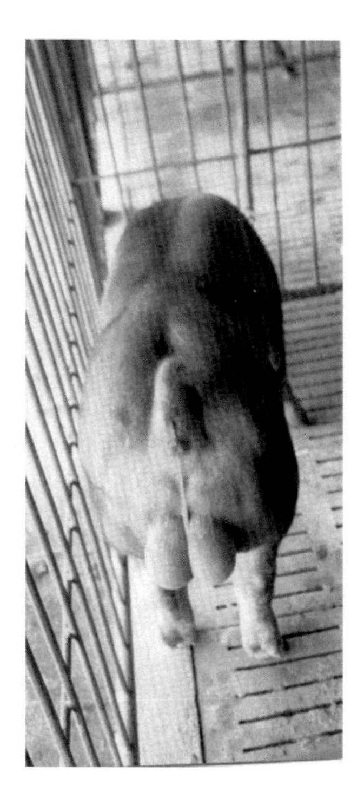

FIG. 8.12

Duroc boar shown in this figure reflects an example of the "butterfly" shape of the loin region of a very muscular pig.

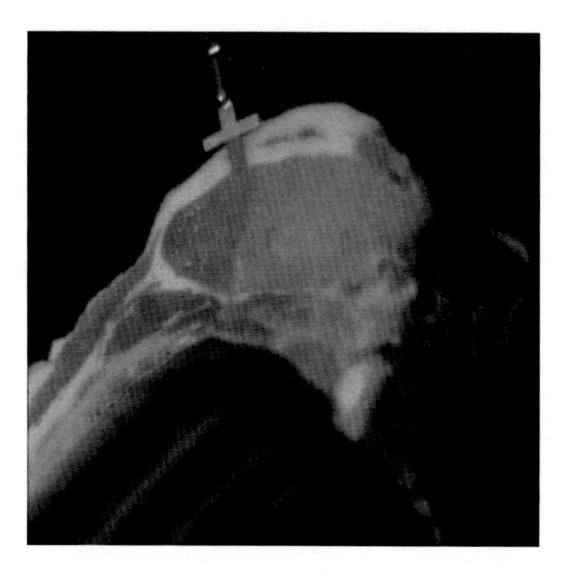

FIG. 8.13

An example of the location of the backfat probe in a muscular and lean carcass from a pig with a butterfly shape of the loin region.

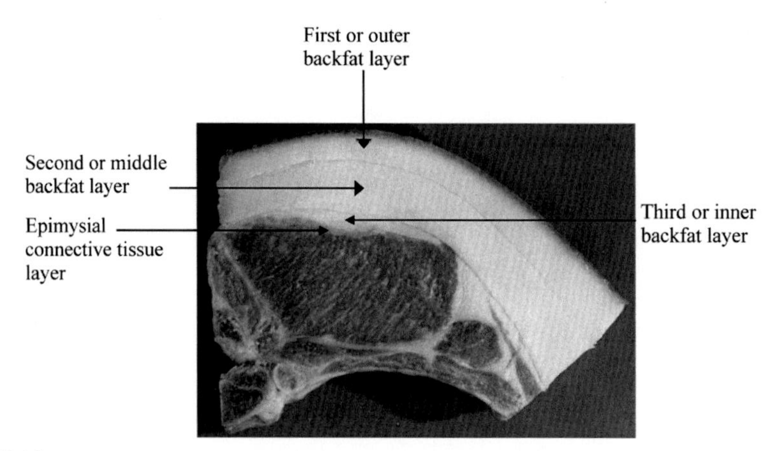

FIG. 8.14

An example of the three layers of backfat separated by connective tissue in the loin region of the pig.

The probe should not penetrate below the third connective tissue layer (Fig. 8.14) because this would give an abnormally high estimate of the fat depth. With some experience, the person probing the pig can easily determine when the epimysial connective tissue located above the loin eye muscle and under the third layer of backfat is reached. To obtain accurate readings, it is also important to understand the locations of the three connective tissue layers when scanning for fat thickness and loin eye area with the real-time ultrasound equipment.

Once the fat depth is known, it can be used in a regression equation with other variables such as live weight and muscling score to estimate carcass composition. Fat depth alone accounts for most of the variation in the percentages of carcass composition, but live weight, loin eye area, and muscling score can improve the accuracy of the prediction equation. An example of a prediction equation to estimate the pounds of fat-free lean in the pork carcass is listed as follows:

$$\text{Pounds of fat} - \text{free lean} = 6.783 - 15.745 \times \text{avg.fat depth from Fat} - \text{O} - \text{Meater} + 4.007 \times \text{avg.muscle depth from Fat} - \text{O} - \text{Meater} + 0.47 \times \text{hot carcass weight.}$$

The major advantages of the backfat probe are cost-effective concepts because it is a good prediction of carcass composition and can be obtained at a very low cost. It is a rapid method and used mostly for selection of breeding animals. If ultrasound equipment is not available, the backfat probe is still a useful method to reduce fat in the selection of breeding pigs and cattle. For lambs and breeding sheep, the variation in fat thickness over the rib section is limited and not large. Therefore good measurements obtained with a ruler-type probe are more difficult to obtain.

Backfat probe for cattle

Beef cattle must be restrained before the fat-probing procedure is started. In cattle, the fat thickness probe is placed through the skin approximately 5 in. from the midline between the 12th and 13th ribs. A modified needle probe is usually used in beef cattle rather than a small ruler that is used in pigs (Fig. 8.15). The needle probe is more effective for penetrating the thick skin of beef cattle and consists of a thick stainless steel wire attached to a metal ruler. The wire of the wire-ruler assembly is inserted through the hub of a hypodermic needle, and the ruler displays the fat probe thickness directly in increments of 0.02 in. Beef cattle also have three layers of subcutaneous fat so the person entering the needle probe in cattle must develop a technique so the aponeurosis connective tissue is penetrated but not the epimysial layer over the longissimus muscle. Other than using a needle probe, the process for probing cattle is similar to the pig, but in cattle one has to adjust for a thicker hide thickness. In beef cattle, the hide thickness can be two times the skin thickness in pigs. Real-time ultrasound data would indicate that there can also be a twofold difference in hide thickness among cattle. Once the fat thickness is recorded, it can also be used for prediction equations to estimate percentage fat and muscle in the beef carcass.

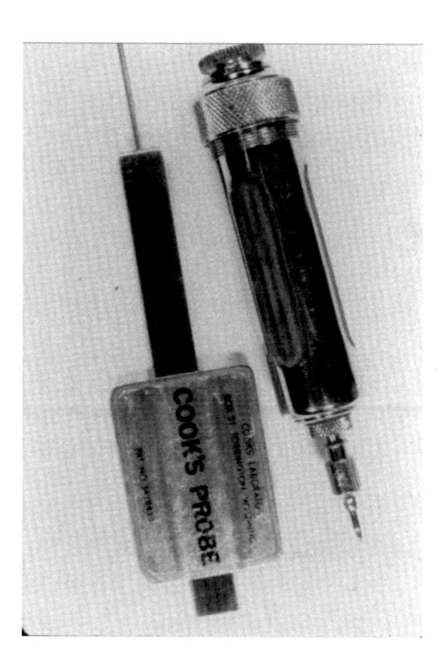

FIG. 8.15

An example of the needle probe used for estimating fat thickness in beef cattle at the 12th–13th rib.

Courtesy of P. Brackelsberg, Iowa State University, Animal Science Department.

Linear measurements used for carcass evaluation

Linear measurements used for pork, beef, and lamb carcass evaluations have been used for simple and inexpensive evaluations methods for more than 50 years, and they are still used today. An inexpensive ruler can be used to measure fat thickness in the carcass, and a measuring tape can be used to determine the length of the carcass. The longissimus muscle area in the carcass can also be estimated by measuring the muscle depth and entering the depth in inches to a prediction equation. Examples for measuring backfat thickness and carcass length in the pork carcass are shown in Fig. 8.7.

ULTRASONIC TECHNOLOGY

Ultrasonic technology used to measure and estimate body composition is the most common objective method used today for both live-animal and carcass evaluations. The method is fast, noninvasive, and it is the most accurate of the commercial methods currently available to the livestock and meat industry.

Concepts for ultrasound technology

Ultrasonics is based on the principle of high-frequency sound waves passing through the animal tissues. A pulse generator sends electrical pulses, and the electrical pulses are converted into sound waves in the transmitter. When an interface between two tissues is encountered, some sound waves are reflected back. The reflected sound waves pass back through the tissues and are absorbed by the receiver and amplified before they are shown in a visual form. Variations in the time taken for the reflected waves to return to the transducer are used to measure variations in the distances of the boundaries between tissues to display the tissue image on a screen. This is the "A-Mode" concept.

Real-time ultrasound

Real-time ultrasound technology is the "B-Mode" concept and is used extensively to evaluate domestic animals for carcass traits, and it is also used by some companies that harvest pigs to estimate muscle percentage in pork carcasses for commercial pricing of pigs. Real-time ultrasound produces an instantaneous picture of the tissues evaluated. The instantaneous image is the result of a linear array of multiple transducers that are stimulated in succession. The action sends sound waves into the tissue. The sound waves interact with the different tissues to form patterns of energy within the tissues. The patterns can be observed and measured on a video monitor. The display on the screen is updated instantaneously to create a two-dimensional image of the anatomical region of the body evaluated.

With the real-time images, fat depth and longissimus muscle depth and muscle area can be measured with a high degree of accuracy and repeatability. An example of the real-time ultrasound machine scanning Angus cattle is shown in Fig. 8.16.

Real-time ultrasound technology can be used to decide when feedlot cattle should be marketed because fat thickness, rib eye area, and marbling can be determined and related to live weight and animal age. With this information, feedlot managers can make good decisions on when to market cattle to obtain the greatest financial returns.

FIG. 8.16

An example of real-time ultrasound scan on Angus cattle.

Courtesy of Dr. Gene Rouse, Iowa State University.

Examples of ultrasound real-time scans are shown in Fig. 8.17. This information can also be used in a genetic selection program. An example is described for beef cattle. Real-time ultrasound images are collected by certified technicians in the United States and internationally. These images are then sent to centralized processing laboratories to be interpreted, and the data are sent to beef breed associations where ultrasound EPD for carcass traits, rib-eye area, fat cover, and percent intramuscular fat are calculated. Currently, all major breed associations calculate and publish carcass EPD on 200,000 yearling bulls, replacement heifers, and market steers per year.

Ultrasound marbling prediction

The ultrasound technology and software systems to predict marbling in pigs and cattle have much to offer for the genetic selection of boars and bulls in a breeding program. Currently, this technology is used primarily for genetic selection programs in cattle. The Angus Certified Beef program is an example where ultrasound predictions of marbling in bulls can be used to select sires for marbling traits. This is an essential part of the Angus Certified Beef program to ensure a good supply of highly marbled beef. Dr. Gene Rouse and Dr. Doyle Wilson, Iowa State University, were pioneers for using real-time ultrasound technology to predict marbling in bulls that were used to improve marbling in their offspring.

REFLECTANCE PROBE FOR PORK CARCASS EVALUATION

The concepts for the reflectance probe (Fat-O-Meater) were developed at the Danish Meat Research Institute. It is sold by SFK Technology, Herlev, Denmark. The reflectance probe measures the difference in light reflectance of fat and muscle tissue. The

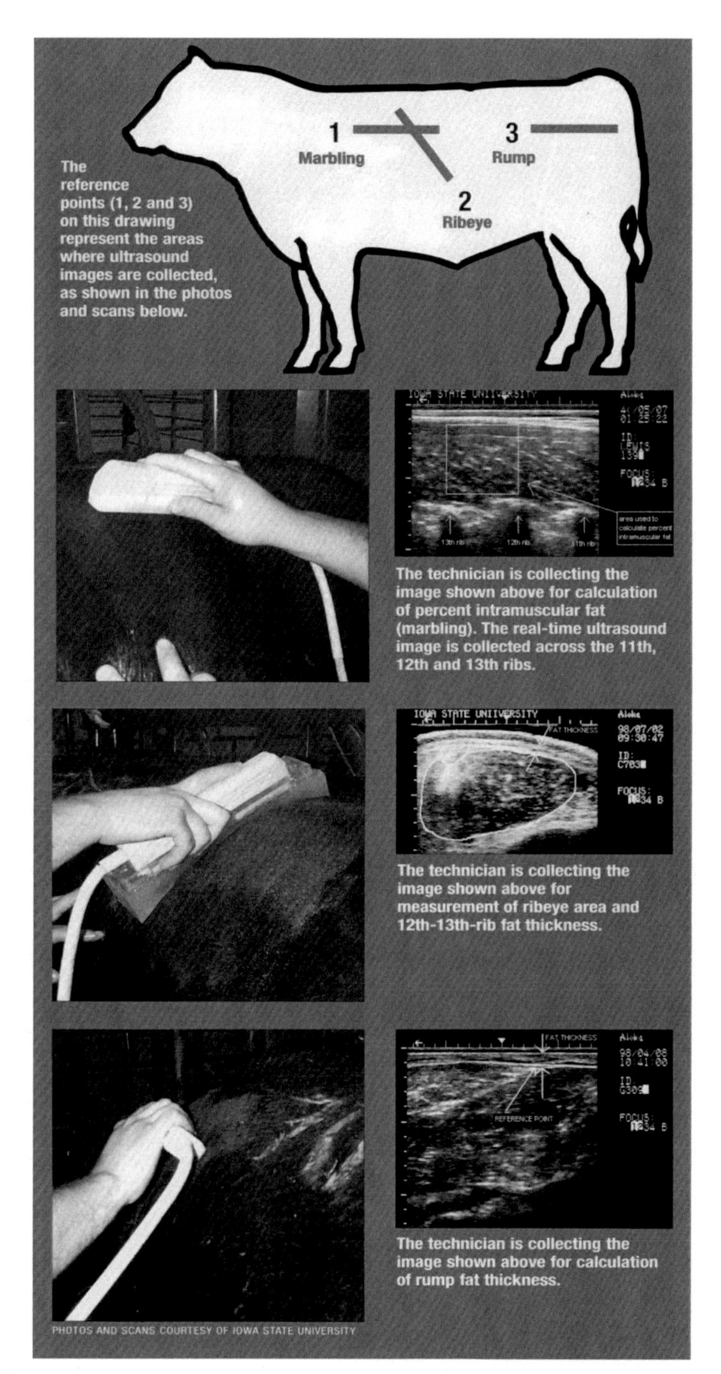

FIG. 8.17

An example of images from an ultrasound machine showing marbling, fat thickness, and rib-eye area.

Courtesy of Dr. Gene Rouse and Dr. Doyle Wilson, Iowa State University, and John Crouch, American Angus Association.

probe records light reflectance every 0.5 mm as it passes through backfat and muscle tissue. An example is shown in Fig. 8.18. The probe is inserted into the 10th to last rib section of the pork carcass and 2 in. off the midline of the carcass. Examples of reflectance probe measurement locations are shown in Fig. 8.19. The capacity of the reflectance probe is about 1200 carcasses per hour. The reflectance data for each carcass can be combined with the carcass weight to predict the carcass muscle percentage.

OTHER METHODS USED FOR PREDICTING CARCASS COMPOSITION

Many techniques for estimating body and carcass composition were developed over the last 50 years, but as less expensive or more accurate methods were established, the older technology found limited use and some methods were only used for

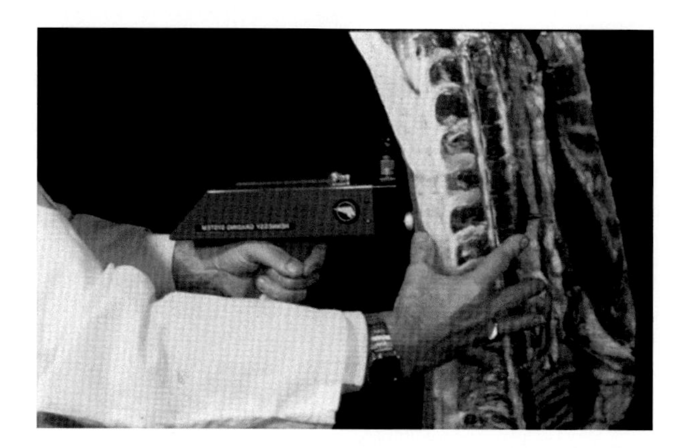

FIG. 8.18

An example of the reflectance probe of Fat-O-Meater for pork carcass evaluation.

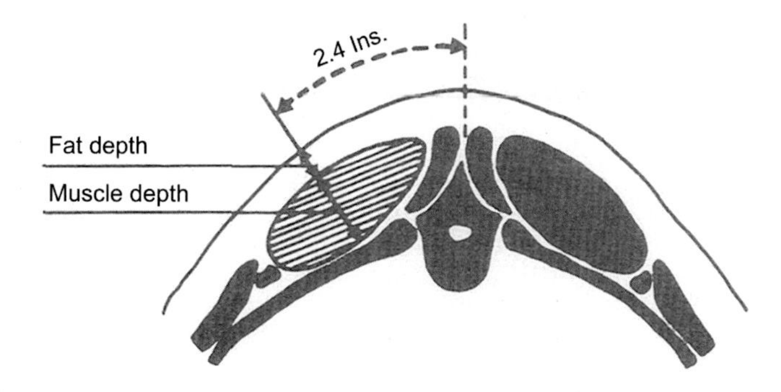

FIG. 8.19

Example of the position for measurements for the Fat-O-Meater or reflectance probe in the loin region of a pork carcass.

research. Some examples of the older techniques are whole-body counters for ^{40}K, video image analysis, body density, electronic meat-measuring equipment (EMME), urea dilution techniques, total body electrical conductivity (TOBEC), and magnetic resonance imaging (MRI). The MRI technology is good but not practical for the animal and meat industry. Recent research, however, showed that quantitative magnetic resonance technology provided a good method to measure fat, lean, and water content of broiler chickens.

SUMMARY

A wide variety of methods to estimate live-animal and carcass composition are available to the livestock and meat industry. Subjective methods such as visual evaluations of live animals were the early methods of choice, and they are still used in today's industry. The visual methods are not as accurate as the objective methods, but they are convenient and not expensive. Livestock shows at the county, state, and even national level still use visual evaluations by expert judges, even though excellent technology for objective measurements exists to estimate live and carcass composition.

Visual methods combined with some objective methods are still used by the USDA Grading System for beef carcasses. Lamb-carcass grading is still based on visual methods. Some objective methods are also low cost and useful for estimating live and carcass composition. An example would be the backfat probe, which is a small stainless steel ruler, but ultrasound technology has replaced the backfat probe. Ultrasound technology is the current technology of choice to estimate live and carcass composition as well as marbling (intramuscular fat) in live animals.

QUESTIONS FOR STUDY AND DISCUSSION TOPICS

1. Describe the conformation or shape of the ham from the back view for a meat-type and fat-type pig.
2. How accurate are highly trained judges when they evaluate pigs, cattle, or sheep by subjective methods (e.g., visual appraisal) to estimate the percentage of muscles or fat in the carcass of the animal?
3. Are the USDA Grading Standards used for lamb and beef carcasses based on both quality and conformation traits and are they determined by subjective (visual) methods?
4. In pork carcasses, what linear measurements are used to evaluate the carcass for muscle or fat percentage?
5. Describe the backfat probe used to predict fat thickness in pigs, and what effect did it have to improve the genetic base for leaner pigs?
6. Describe the real-time ultrasound concepts required to obtain an instantaneous image of the backfat thickness and loin-eye area in a market-weight pig.
7. Can ultrasound technology be used to predict marbling in cattle and pigs?

CHAPTER

Intrinsic cues of fresh meat quality

9

INTRODUCTION

As a commodity, fresh meat is most commonly characterized by proximate composition. Fresh meat cuts are frequently categorized by composition, specifically with regard to intramuscular lipid or marbling. This is especially true with beef, as the marbling content determines the quality grade and value of fresh beef. We often refer to the materials we use in further processing as 90/10, 80/20, 50/50 to describe the lean/fat ratio in the raw material. There are more specific properties of fresh meat that have a direct contribution to how we use raw materials in further processing or their application as a fresh meat item. The precise structure and chemistry of muscle and meat have direct influences on the property of the raw material. Many times, this is dictated by the function of the muscle in the animal. The function of the muscle in the animal may direct the type of energy stores, the muscle fiber type, or the required amount of connective tissue. The profile of these properties contributes to the properties of fresh meat including water-holding capacity, color, and tenderness. This chapter will specifically focus on these traits and the intrinsic properties of meat from skeletal muscle that contribute to the variation of these important properties.

WATER-HOLDING CAPACITY

The ability of fresh meat to bind water (water-holding capacity) is an important quality feature. Variation in water-holding capacity can create inconsistencies in fresh meat that influence processed products. The ability of meat to bind and hold water can vary a great deal. The consequences of water lost as drip or purge from fresh meat is loss of weight, loss of protein, and loss of water-soluble vitamins. It is also recognized that meat products that have poor water-holding capacity during storage will often exhibit poor sensory quality.

In processed meat, the primary added nonmeat ingredient is water. Thus the ability to bind added water is of critical importance to the success of processing to add value.

The ability of fresh meat to bind water is attributed to proteins and their structures within muscle. Specifically, *myofibrillar proteins* bind water much more effectively than stromal or sarcoplasmic proteins. *Myofibrils* account for almost 80% of the

The Science of Animal Growth and Meat Technology. https://doi.org/10.1016/B978-0-12-815277-5.00009-3

volume of muscle cells. It is therefore not surprising that much of the water in fresh meat is held within the myofibrils.

NET CHARGE EFFECT

Proteins are *amphoteric* molecules. They can be an acid or a base, and this is specifically dependent on the pH of their environment (Fig. 9.1). This characteristic of proteins is principally due to the side chain groups of the constituent amino acids. Seven of the 20 amino acids contain side chains that are readily ionizable. The approximate pK_a of each of these is provided in Table 9.1. It is clear that amino acid composition can have a large effect on the amphoteric character of proteins.

The ability of proteins to bind water directly is mostly due to the functional side chain groups of constituent amino acids that give proteins a charge. The *net charge effect* relates to the net charge on a particular protein. The isoelectric point (p*I*) of a protein is the pH at which a specific protein has a net charge of zero. In the example in Fig. 9.1, the molecule depicted has a p*I* of 6.0. Note that this does not mean there is no charge. Rather, the positive and negative charges are equivalent. A protein that is exposed to a pH above the p*I* will have a net negative charge. A protein that is exposed to a pH below the p*I* will have a net positive charge.

Water molecules do not specifically have a charge, but they are dipolar. The dipole nature of water contributes to the interaction with polar groups on proteins. Thus

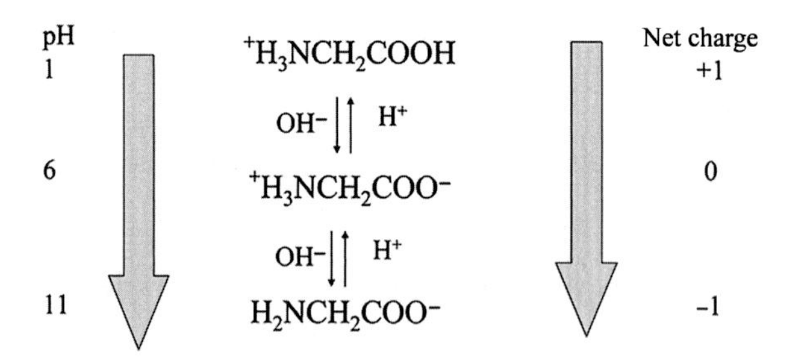

FIG. 9.1

Example of the effect of pH on net charge of a molecule.

Table 9.1 Ionizable side chains of amino acids in proteins

Group	Approximate pK_a
Aspartic acid	4.4
Glutamic acid	4.4
Histidine	6.5
Cysteine	8.5
Tyrosine	10.0
Lysine	10.0
Arginine	12.0

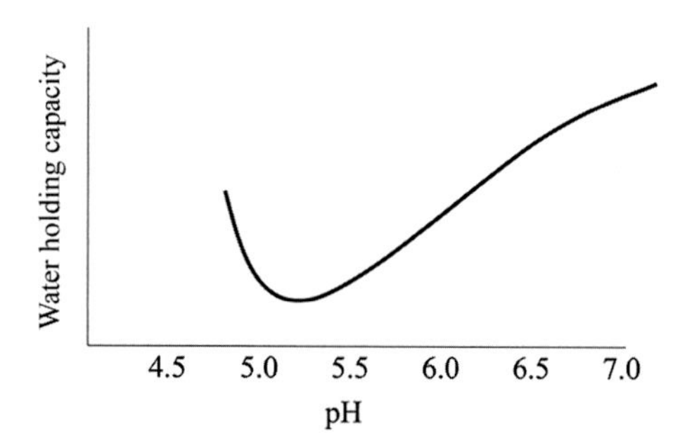

FIG. 9.2

Effect of pH on water-holding capacity of fresh meat.

proteins with greater net charge will tend to bind more water. The collective p*I* of the major myofibrillar proteins is about 5.2. Therefore the poorest water-holding capacity will be expected at pH 5.2 (Fig. 9.2). As pH increases, the net charge increases and thus allows more direct protein-water interactions. As pH decreases the net positive charge increases and thus direct water and protein interactions also increase. However, other factors, specifically protein denaturation at very low pH, can minimize this effect.

STERIC EFFECTS

Although the ability of proteins to bind water directly is absolutely essential, it is estimated that only 5% of the water within a muscle cell or within meat could be bound directly through the hydrophilic interactions that were just described. In fresh meat, the great proportion of the rest of the water is held within the myofibrils by capillary forces. It is thus important to recognize that the volume of the myofibril is key in determining water-holding capacity. Much of the water bound within the myofibril resides between the thick and thin filaments (Fig. 9.3). The distance between these two filaments can vary from 320 to 570 Å and is dependent on the pH, ionic strength, and whether the muscle is pre- or postrigor. This change in filament spacing represents a very large difference in myofibrillar volume. It is of no surprise that formation of the rigor bonds can decrease the interfilament spacing.

The steric effects are not entirely independent of the net charge effects. In general, greater pH and ionic strength increase interfilament spacing (Fig. 9.4). The pH and ionic strength effects are linked to the net charge effect, because protein filaments with net negative charge are likely to generate some electrostatic repulsion and increase in myofibrillar volume.

EFFECT OF PROTEOLYSIS

It is important to remember that a great deal of the water (up to 80%) is held within the myofibrils and muscle fibers in muscle. Loss of water from the intracellular space creates "drip channels" along the endomysial and perimysial layers. Therefore maintenance of water within the cell is of importance to improve water-holding capacity.

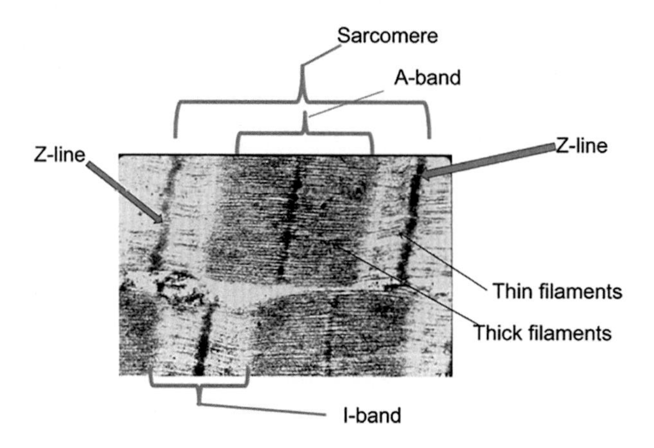

FIG. 9.3

Electron micrograph of a myofibril from skeletal muscle. The sarcomere is the basic contractile unit of skeletal muscle. It spans from Z-line to Z-line. Thin filaments anchored in the Z-line slide past the thick filaments in the A-band to allow muscle contraction.

Courtesy of Iowa State University Animal Science Department; M. Stromer and D.G. Goll.

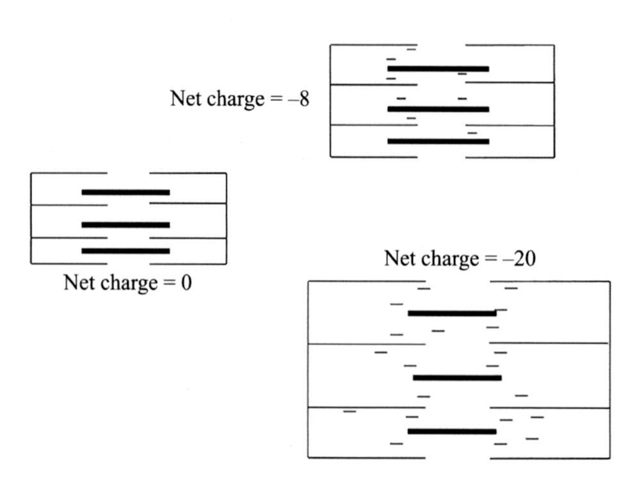

FIG. 9.4

Effect of net charge on myofibrillar swelling. Increasing the net negative charge on the thick and thin filaments will result in myofibrillar swelling and a greater opportunity to trap and hold water within the myofibril.

As muscle is converted to meat, the pH decline will obviously allow some shrinkage of the myofibrils. This is due in part to the net charge effect but also due to some protein denaturation. If connections between myofibrils and between the myofibrils and sarcolemma are intact, the shrinkage of the myofibrils will translate into shrinkage of the muscle fiber and loss of water to the drip channels. If, however, the linkages are broken, the shrinkage of the myofibril (which is inevitable) does not result in shrinkage of the muscle cell. It stands to reason then that degradation of the linkages between the myofibrils and between the myofibrils and the sarcolemma will result in better water-holding capacity in fresh meat.

Of course, the significant protein linking the myofibrils and the sarcolemma is the intermediate filament protein desmin. Desmin is a known substrate of the calpain proteinase, μ-calpain. There is a known association between desmin degradation and improved water-holding capacity in fresh meat. Conditions that contribute to increased proteolysis in early postmortem muscle tend to result in improved fresh meat water-holding capacity.

FRESH MEAT COLOR

Fresh meat color is an important quality feature. To an average consumer, meat color is the first sensory experience of the product. Most consumers expect to see a pleasing bright red color of fresh meat. Variations from the expected color cause reduced sales at the retail counter because consumers do not like to purchase meat when they observe poor quality.

To the trained food professional (meat buyer, meat processor, chef, etc.), meat color is more than just the first impression of the product. Meat color is an indicator of the type of muscle the meat cut comes from, can indicate the relative age of the animal that produced the cut, can be a predictor of other traits such as water-holding capacity and tenderness, and can also indicate the packaging history and shelf life of the meat product. This section will focus on the chemistry behind meat color.

Color is a sensory trait in the sense that our senses experience color. As sensed by the eye, color is a combination of *dominant wavelength* (*or hue*), *chroma*, and *value*. Hue is what we normally think of as color (red, green, blue, yellow). It is the dominant wavelength of the light radiation that is reflected. Chroma indicates the purity of the color. It represents the proportion of the total amount of light reflected that is at the dominant wavelength. Value represents the light reflectance of a color.

Color of meat products can be evaluated in several different ways. In general, trained meat evaluation professionals or meat graders use their experience and subjective evaluation to evaluate meat products based on differences in color of the product. Instruments are also used to distinguish differences in hue, chroma, and value. Common instruments used in the meat industry measure color reflectance and report L values [for lightness of the product (0 is black, 100 is pure white)], a values (positive values indicate intensity of red light reflectance, negative values indicated intensity of green light reflectance), and b values (positive values for intensity of yellow and negative values for intensity of blue light reflectance). Fig. 9.5 is a depiction of the L, a, b rectangular color space. Typical L, a, and b values for beef, pork, and chicken are provided in Table 9.2.

Meat color is primarily influenced by the pigment protein myoglobin. Myoglobin is classified as a metalloprotein because it has a protein portion made up of an amino acid chain covalently linked to an organic iron structure called heme iron. The function

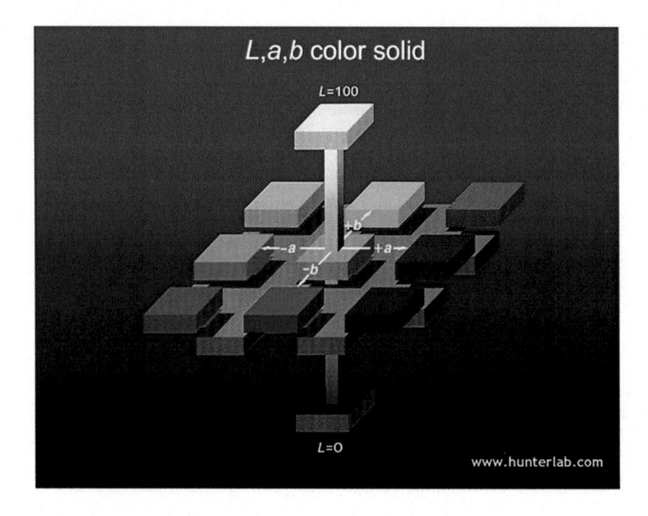

FIG. 9.5

Depiction of *L*, *a*, *b* rectangular color space.

Courtesy of HunterLab.

Table 9.2 Color parameters (L^*, a^*, b^*) of fresh beef, pork, and chicken

Species	Muscle	L^*	a^*	b^*
Beef[a]	Longissimus dorsi	40.6	31.13	23.98
	Psoas major	34.1	27.2	20.9
Pork[b]	Longissimus dorsi	46.4	3.7	10.3
	Psoas major	43.04	10.9	15.01
Chicken[c]	Pectoralis major	51.8	1.5	8.8
	Biceps femoris	49.1	2.7	9.5

[a] *Data are from a bovine myology database (http://bovine.unl.edu/).*
[b] *Data are from Melody et al., 2004, Journal of Animal Science, vol. 82, p. 1195.*
[c] *Data are from Lonergan laboratory (unpublished).*

of myoglobin in muscle is to serve as an oxygen storage molecule for oxidative metabolism. It stands to reason that muscle fibers that depend on oxidative metabolism for energy (Type I fibers) have a higher concentration of myoglobin than do muscle fibers that have a greater dependency on glycolytic metabolism (Type II fibers).

Other proteins—such as cytochrome proteins—contribute to meat color but to a much lesser extent than myoglobin. The *amount of myoglobin* has obvious effects on meat color. This might be best envisioned by comparison of a chicken breast to a beef rib eye steak, with the rib eye steak having a 7–10 times greater concentration of myoglobin. Within species, muscles from older animals have a greater myoglobin

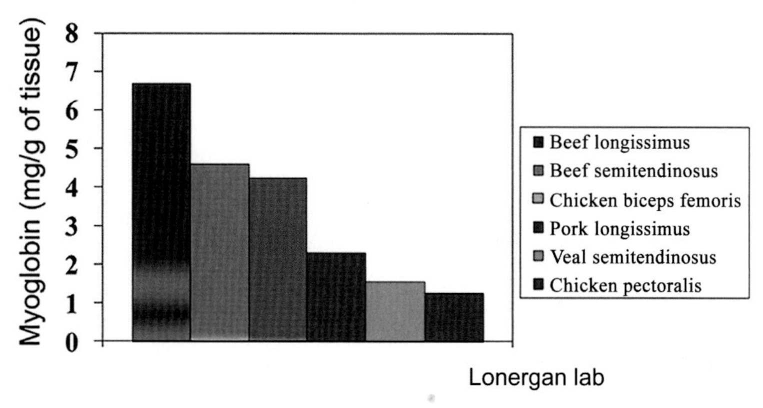

FIG. 9.6

Myoglobin content of different muscles and species.

concentration than the same muscles from younger animals (i.e., beef vs veal). As stated earlier the metabolic basis (glycolysis vs oxidative) influences the myoglobin content. Chicken breast is from the pectoral muscle that is made up of almost completely glycolytic fibers. Chicken thigh muscles are a mix of glycolytic and oxidative fibers, thus explaining the darker color of those cuts (Fig. 9.6).

A separate issue that affects myoglobin concentration is changes in myoglobin solubility. Remember that conditions that contribute to development of pale, soft, and exudative pork (high temperature and low pH) denature many proteins, including myoglobin. Denatured myoglobin does not contribute to the red color of meat. Any process that results in a change in myoglobin solubility therefore has the potential to directly affect the color of fresh meat. The change in color of a ground beef patty after it is placed on a skillet for only a few seconds (from bright red to gray) is caused by denaturation of myoglobin and other proteins.

The *oxidation state of the heme iron* directly influences the color reflected by myoglobin (see Fig. 9.7). When the iron is in the ferrous state (Fe^{2+}), the molecule has a high affinity for oxygen. If oxygen is present, the myoglobin forms *oxymyoglobin*. This is the form of myoglobin that is responsible for the bright red color of beef and pork that is observed in traditional retail meat packages with oxygen-permeable wraps. If oxygen is not available and the ferrous iron is not bound to gas, the form of myoglobin is *native myoglobin* or *deoxymyoglobin*. When a great proportion of myoglobin is present as deoxymyoglobin, the meat is typically a darker reddish purple color. This color is observed immediately after fabrication of primal and subprimal cuts. As cuts are exposed to open air the color "*blooms*" from the deoxymyoglobin deep reddish purple to a brighter red color imparted by the oxymyoglobin. The same phenomenon can be observed when one opens a vacuum-packaged fresh beef steak and allows conversion from deoxymyoglobin to oxymyoglobin.

Oxidation of the heme iron results in a transition from the ferrous iron (Fe^{2+}) to ferric iron (Fe^{3+}). Ferric iron does not have a high affinity for oxygen. In fact, it likely

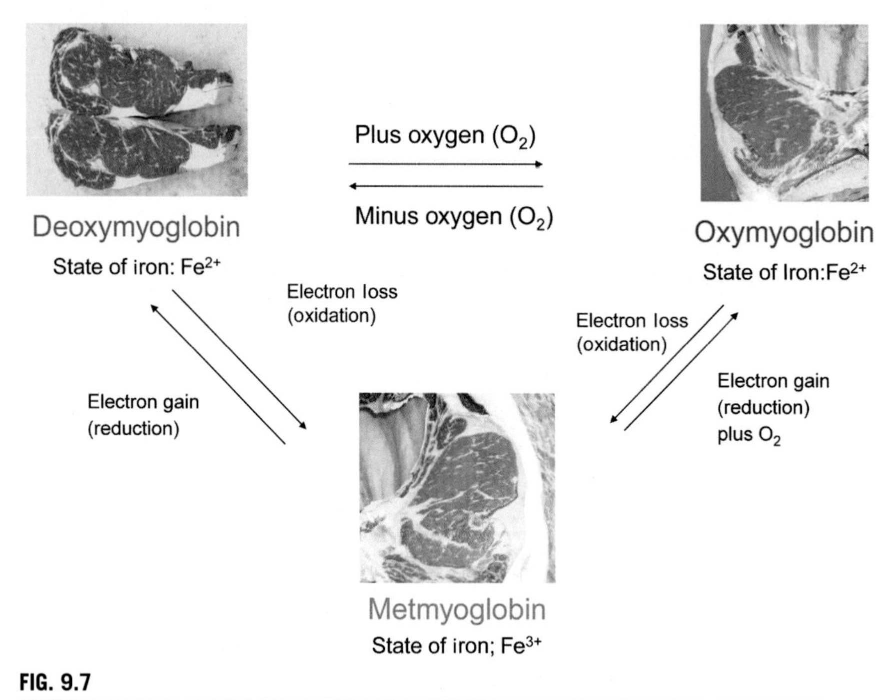

FIG. 9.7

Classic fresh meat color triangle.

binds water vapor. This transition is definitely negative for the appearance of the product. Oxidized pigment—*metmyoglobin*—imparts a gray or grayish brown color to meat. Meat that has been exposed to the environment for a long period of time will demonstrate development of this oxidized pigment. Oxidation of the pigment is catalyzed by temperature abuse. In general, consumers are correct in avoiding discolored meat in the retail case.

Oxidation of myoglobin does represent a significant loss of product value at the retail level. If product discolors due to oxidation, sales are lost. Improving color shelf life is of utmost importance to meat retailers. One approach is to elevate levels of antioxidants in muscle to extend color shelf life. This can be achieved by adding ingredients. However, an efficacious approach is to feed high levels of vitamin E (α-tocopherol) to cattle during the finishing phase, usually during the last 100 days of feeding. The accumulation of this fat-soluble antioxidant appears to be responsible for slowing down the accumulation of the oxidized pigment, metmyoglobin.

Although myoglobin concentration, solubility, and oxidation state have a profound effect on fresh meat color, the physical structure of meat also influences the way light reflects from meat. Dark-cutting beef or dark, firm, and dry pork both have dark colors that are not specifically related to pigment concentration. Remember that these conditions arise from high ultimate pH in meat and are typically caused by depletion of glycogen before slaughter (Chapters 4 and 10). As reviewed in this chapter, high pH results in very high water-holding capacity. In fact, the myofibrils within each muscle cell bind so much water that the myofibrils swell to the point of packing each muscle cell and muscle bundle. When this occurs, light is readily absorbed and not reflected. This is the explanation for the dark color. In contrast, pale, soft, and exudative meat (Chapters 4 and 10) has a very loose structure because myofibrils do not hold water very efficiently under these conditions. The loose structure allows more reflectance of light and the appearance of a lighter cut of meat.

TENDERNESS

Tenderness is a key feature of fresh meat. It might be argued that beef tenderness is only one component of beef palatability. Indeed, beef juiciness and flavor also affect consumer acceptability. However, research repeatedly demonstrates that (1) consumers can detect differences in tenderness, (2) tenderness classifications affect overall satisfaction based on sensory panels, and (3) consumers are willing to pay a premium for meat they are confident is tender.

Tenderness of all meat products is important, but most of the time the focus is on whole-muscle products. This is likely because ground products are less variable in overall texture. Tenderness of cooked whole-muscle products can be measured in several ways. Many instrumental methods exist, although the most commonly used in research and product development is the Warner-Bratzler shear. This method measures the force necessary to shear through a half-inch core of cooked product with a dull blade. Other instrumental methods that are used include a star probe (measurement of force necessary to compress a cooked chop or steak) and a slice shear (measurement of the force necessary to shear through a slice of a chop or a steak). Sensory tenderness can also be measured using a sensory approach. This approach uses human subjects to describe or quantify a trait such as tenderness, texture, juiciness, and so on. A trained sensory panel is usually a small group of individuals who are constantly refining their precision and repeatability of evaluation of particular traits. Consumer panels are also useful in that product researchers can monitor differences in products that consumers can detect. In contrast to the trained panels, consumer panels usually require hundreds of observations. Regardless of the method chosen, instrumental measurement of cooked meat texture requires a great deal of precision and attention to detail.

So, what affects tenderness of fresh meat? Many *extrinsic* factors can influence meat tenderness. These include degree of doneness, cutting style, and type of cookery. The *intrinsic* factors that affect fresh meat tenderness include

Sarcomere length,
Connective tissue,
Lipid content (marbling), and
Myofibrillar fragmentation.

Table 9.3 Sensory tenderness, sarcomere length, and collagen content of pork cuts from selected muscles

Muscle	Collagen content (mg/g)	Sarcomere length (μm)	Sensory tenderness[a]
Semitendinosus	5.3	2.45	7.2
Triceps brachii	6.0	2.44	7.1
Longissimus	4.1	1.78	6.4
Semimembranosus	4.5	1.83	5.7
Biceps femoris	7.1	1.74	4.0

[a] *Scores: 1 = extremely tough, 8 = extremely tender.*
From Wheeler et al., 2000, Journal of Animal Science, vol. 78, pp. 958–965.

Typically all four of these features come into play to contribute to the tenderness of fresh meat. Longer sarcomeres indicate less opportunity for thin and thick filament overlap and decrease the number of actomyosin cross-bridges, and thus there is a more tender product. Greater connective tissue content and connective tissue insolubility result in a less tender product. In general the cut with the greatest connective tissue content and the shortest sarcomere length (biceps femoris) was determined to be the least tender by a trained sensory panel (Table 9.3). Intramuscular lipid content (marbling) is associated with somewhat more tender product. Finally, more postmortem myofibrillar fragmentation is known to improve meat tenderness.

SARCOMERE LENGTH

There is good evidence that a great deal of the toughening that occurs early postmortem is the result of rigor shortening (Table 9.3). Indeed, using a laboratory application, it has been demonstrated that if shortening is prevented, meat toughening does not occur. In the contrasting situation—cold shortening (discussed in Chapter 10)—uncontrolled shortening caused by release of calcium before depletion of ATP results in very short sarcomere length and product toughening. Some degree of rigor shortening always occurs, but it is rarely as extreme as observed in cold shortening. Table 9.4 contains data that demonstrate the change in sarcomere length observed early postmortem before and immediately after the onset of rigor mortis. It is no surprise that as the sarcomere length decreases, the shear force increases.

CONNECTIVE TISSUE

Connective tissue is an integral part of muscle structure. The endomysium, perimysium, and epimysium form the connective tissue network around the fiber, bundle, and muscle (Fig. 9.8). Muscles that are used for locomotion tend to have a greater proportion of connective tissue. The primary connective tissue protein is collagen. Collagen content is often measured in meat to get an estimate of the amount of connective tissue.

Table 9.4 Sarcomere length changes during the early postmortem period

Time postmortem (h)	Sarcomere length (μm)	Warner-Bratzler shear force (kg)
0	2.24	5.07
3	2.00	5.10
6	1.80	6.53
9	1.72	8.26
12	1.75	8.24
24	1.69	8.66

From Koohmaraie et al., 1996, Journal of Animal Science, vol. 74, p. 2935.

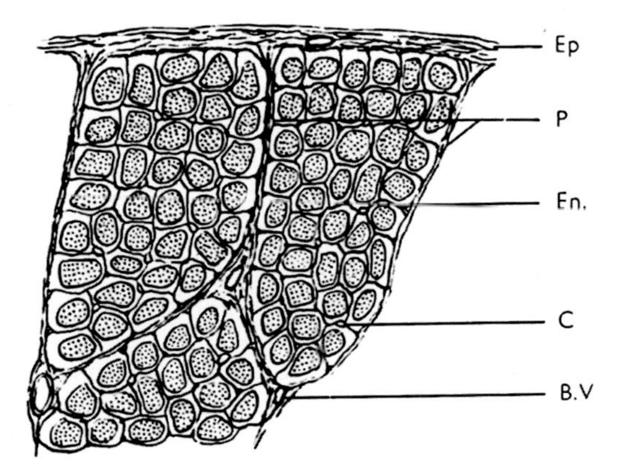

FIG. 9.8

Drawing of a cross section of muscle showing the associated tissues with muscle. *Ep*, epimysium—the connective tissue layer surrounding the entire muscle; *P*, perimysium—the connective tissue layer surrounding the muscle bundles; *En*, endomysium—the connective tissue layer surrounding the muscle fiber; *C*, capillary; *B.V.*, blood vessel.

Courtesy of Iowa State University Animal Science Department; M. Stromer and D.G. Goll.

The amount of collagen is a reflection of the muscle function. Muscles that are used for locomotion typically have a greater collagen content than do muscles that are used for fine control or even posture. Examples of muscles with relatively low collagen concentrations are the psoas major and the longissimus thoracis. Cuts from these muscles are considered very tender regardless of cookery.

In general, large muscles from the limbs (triceps brachii, serratus ventralis, semimembranosus, biceps femoris) have greater amounts of connective tissue and are typically less tender when prepared by dry heat or rapid heating. Table 9.3 summarizes data to show that greater collagen content in different muscles can result in less tender meat. Cookery that uses lower temperature and moist environments are conducive to "melting" collagen and thus results in very tender finished product even when collagen content is great. An excellent example of this is the pectoralis profundi (or brisket). This cut contains a great deal of collagen and connective tissue, but when prepared properly with a low temperature in a high moisture environment, it will have a very soft texture.

MARBLING

Marbling is the visible lipid present in the intrafascicular spaces of a muscle (see Chapter 5, Figs. 5.14 and 5.15). This lipid consists primarily of triglycerides and is associated with the perimysium. Although there is considerable variation in lipid content, intramuscular fat increases with increasing age, decreasing levels of activity, and increasing energy consumption. Intramuscular lipid contributes to the flavor and juiciness and, to some extent, tenderness of meat products. This is useful because it is visible and can be used to classify cuts and carcasses. Because USDA beef grading standards use degree of marbling as a major parameter for US Quality grades, intramuscular lipid content is an important value-defining parameter for beef carcasses.

The contribution of marbling to fresh meat tenderness is of interest to the mcat industry because product specifications for grades and eligibility for export markets or other high-end markets often include marbling. It is clear that marbling does contribute to sensory quality. The questions are often "How much is enough?" or "How much is not enough?" A general observation is that a marbling score of Modest is sufficient to ensure a good eating experience. An increase in marbling score from Modest does not specifically indicate that eating quality will continue to increase. A marbling score of less than Slight will be very likely to produce a product that is lacking in juiciness, tenderness, and overall acceptability.

Pork flavor, juiciness, and tenderness are also influenced by intramuscular lipid content. The influence of intramuscular lipid on objective measures of tenderness in pork is not consistent. Like the general observations for beef, it is commonly held that there is a threshold lipid content for quality of fresh pork. This is thought to be about 2%–3%. An increase in lipid from 3% may not specifically improve pork tenderness. However, lipid content less than 2% may increase the likelihood of a poor quality product.

MYOFIBRILLAR FRAGMENTATION

A common observation is that fresh meat tenderness improves with aging (Fig. 9.9). This is one of the reasons that duration of aging is defined as a specification for high quality fresh meat. As meat ages, structural changes occur in the myofibril (the contractile organelle in the muscle cell). One of the changes that occurs during this time is breakage or structural damage near the Z-line in the I-band (Fig. 9.10). Because of the breakage that occurs in the myofibril and subsequent disruption of muscle integrity, there exists a greater number of myofibrillar fragments in aged than in unaged meat. This increase in the fragmentation of the myofibril has been used to characterize the aging and amount of tenderization that has taken place. Early attempts to use this information have involved procedures such as counting the number

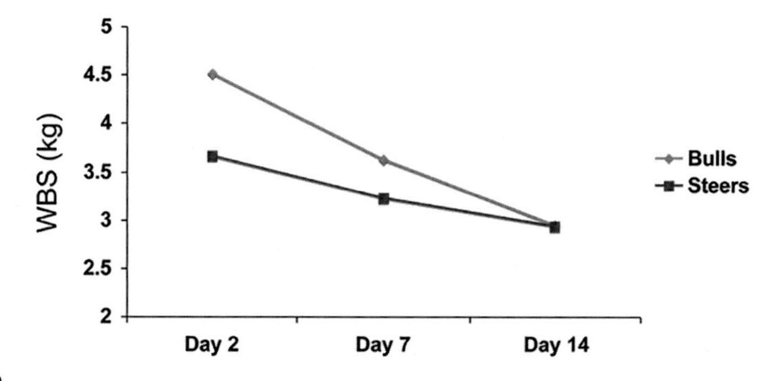

FIG. 9.9

Aging response in beef from bulls and steers. *WBS*, Warner-Bratzler shear force.

From Lonergan et al., 2001, Journal of Muscle Foods, *vol. 12, 121–136.*

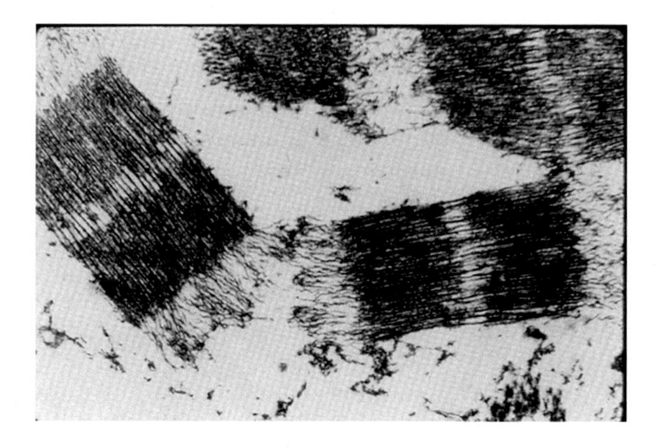

FIG. 9.10

Electron micrograph that depicts the fragmentation of myofibrils that occurs near the Z-line during postmortem aging of meat.

Courtesy of Iowa State University Animal Science Department; M. Stromer and D.G. Goll.

of myofibrils with one to four sarcomeres. This procedure is based on the fact that as postmortem aging time increases, so do the number of shortened myofibrils. Many of the structural changes that lead to myofibrillar fragmentation within the first 72- to 96-h postmortem are thought most likely to be caused by the endogenous, calcium-dependent proteinase, μ-calpain.

Calpains are endogenous cysteine proteases that have an absolute requirement for calcium in vitro to initiate full activity. The two most well-characterized forms of the enzyme are μ- and m-calpain. μ-Calpain has a pH optima near pH 7, but it does retain activity at pH 5.6. The optimum temperature for this proteinase is between 77°F and 86°F, but it does have some activity at refrigeration temperatures. Under conditions of pH 5.5–5.8 and 40°F, it has been shown that μ-calpain retains as much as 28% of the activity that it has at pH 7.5 and 77°F. This points out that μ-calpain may be active under conditions found in postmortem muscle. μ-Calpain has been shown to cause many proteolytic changes in myofibrillar proteins at low pH and temperature. Some documented changes include the release of α-actinin from the myofibril; degradation of titin, nebulin, filamin, desmin, dystrophin, and vinculin; the disappearance of intact troponin-T; and the appearance of polypeptides migrating at approximately 30,000 Da. Degradation of specific myofibrillar proteins is thought to be instrumental in influencing postmortem tenderization of beef. In fact, just as the structure and integrity of the muscle cell do not depend solely on one protein, it is likely that degradation of several specific, myofibrillar proteins by μ-calpain may weaken key areas of the myofibril and the muscle fiber causing an overall weakening of the muscle fiber as time postmortem increases.

A common method used to determine changes in integrity of specific proteins in postmortem meat is called Western blotting. This method uses SDS-PAGE to denature and fractionate the polypeptides based on size, transferring the fractionated proteins to a membrane and detecting specific proteins on the membrane with antibodies. Proteins that were degraded will migrate further down the gel and thus demonstrate protein degradation. Greater protein degradation is indicated by disappearance of the intact protein or appearance of a degradation product. An example of this type of assay is shown in Fig. 9.11. The antibody is specific for troponin-T, a known substrate of calpain and a protein that is degraded in postmortem muscle. When more protein is degraded, the intensity of the 30-kDa degradation product of troponin-T is greater (e.g., in classification 4 in Fig. 9.12A). When this is the case, the product is more tender, indicated by a lower Warner-Bratzler shear force value (Fig. 9.12B). When very little proteolysis has occurred—for example, in group 0—the product will be less tender.

A complicating factor is that the endogenous-specific inhibitor of the calpains, calpastatin, is also present in muscle, and it also retains activity under conditions found in meat. The activity of calpastatin is frequently linked to variation in myofibrillar protein degradation, myofibrillar fragmentation, and development of tenderness in fresh meat. High calpastatin activity in postrigor meat predicts high shear force values and less tender fresh meat.

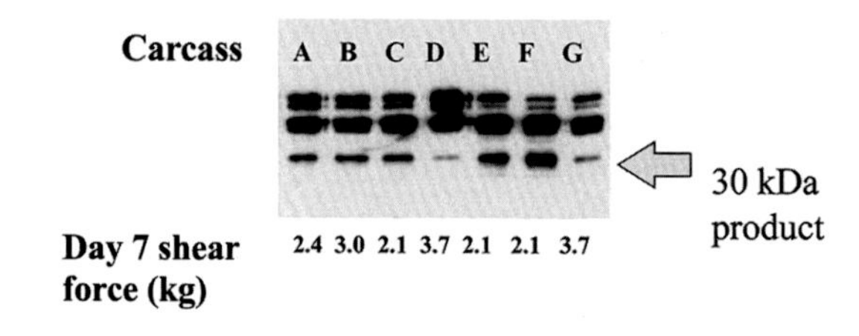

FIG. 9.11

Protein degradation and Warner-Bratzler shear force in beef top-loin steaks.

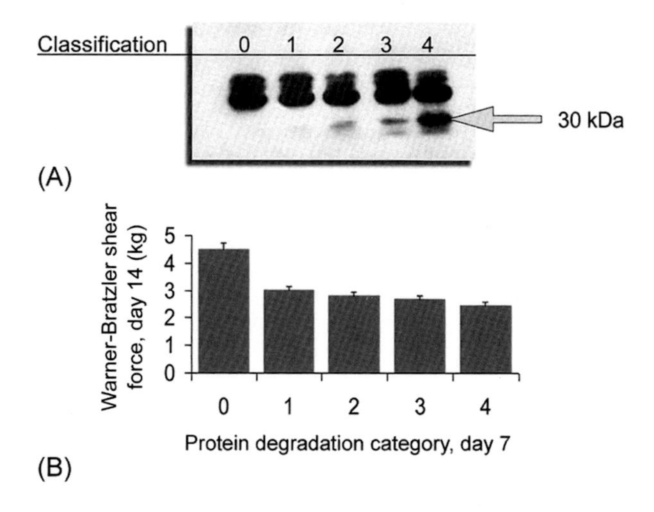

FIG. 9.12

(A) Western blot demonstrating proteolysis differences in different classifications of beef strip steaks. Class 4 demonstrates the greatest degree of proteolysis. (B) Class 4 also has the lowest shear force values. Warner-Bratzler shear force in each protein degradation category.

Courtesy of the Lonergan Lab, Iowa State University.

SUMMARY

This chapter provides examples of how biochemical and structural characteristics of muscle and meat contribute to the properties of meat as food. It is important to note that the function of muscle directly affects the quality and characteristics of fresh meat. Any change in muscle structure or metabolism, therefore has the potential to influence fresh meat quality. It is important to recognize that meat quality is influenced by genetics, growth, and composition of livestock. Furthermore, livestock handling and processing early postmortem all have a significant influence on the intrinsic features of fresh meat that can have direct and profound effects on how consumers perceive meat as food.

QUESTIONS FOR STUDY AND DISCUSSION TOPICS

1. Describe the basis of water-holding capacity using the following principles: Net charge effect and steric effect.
2. Describe how proteolysis of intermediate filament proteins can influence water-holding capacity of fresh meat.
3. How do *RN* or *HAL* genes (from Chapter 10) affect water-holding capacity of fresh meat considering the net charge effect and the steric effects described in this chapter?

4. Sometimes pork is harvested from carcasses before chilling. This product is salted, ground, and frozen. Given your understanding of properties of fresh meat, what do you predict regarding the color and water-holding capacity of this product?
5. Given the features that influence meat tenderness described in this chapter, describe the circumstances where each one might have the most effect on fresh meat tenderness.
6. Provide definitions for the following:
   Calpain
   Calpastatin
   Deoxymyoglobin
   Ferric iron
   Ferrous iron
   Hunter *L*, *a*, *b* values
   Metmyoglobin
   Oxymyoglobin
   Warner-Bratzler shear force

CHAPTER

Conversion of muscle to meat 10

INTRODUCTION

Many dramatic physical and chemical changes occur in muscle when it is converted to meat. These postmortem changes certainly influence the meat quality characteristics of color, tenderness, and water-holding capacity. Indeed, meat quality is associated with the proper care of domestic animals. Hence, proper production systems, as well as postmortem time and temperature practices, are essential and significant in the production of high-quality meat, which results in greater profitability for the farmer, processor, and retail industry. Moreover, the provision of safe, nutritious, attractive, and delicious meat products for the consumer is the ultimate goal of the livestock and meat industry.

PERIMORTEM EVENTS

The condition of the muscle in the perimortem period (in the days and hours before slaughter) sets the stage for the changes that occur in postmortem muscle. The primary reason for this is that the energy status of the muscle before exsanguination dictates the rate and extent of lactic acid production in meat and the onset of rigor mortis. Although hundreds of metabolic events are occurring in the living muscle before slaughter, the primary consideration that influences conversion of muscle to meat is the amount of energy in the form of glycogen. Glycogen is converted to lactic acid in the anaerobic metabolic pathways. The rate and extent of lactic acid production have dramatic influences on protein function. Depletion of energy (specifically ATP) in postmortem muscle leads to rigor mortis (formation of permanent actomyosin cross-bridges and loss of extensibility). The following sections will address the chemical and physical changes in postmortem muscle and will explore how the perimortem environment can affect the progression of these changes.

PHYSICAL CHANGES IN POSTMORTEM MUSCLE

Physical changes in postmortem muscle are characterized by three stages or phases of rigor mortis manifested in skeletal muscle. These three stages or phases are termed delay, onset, and completion of rigor mortis (Fig. 10.1). As long as resting ATP levels

The Science of Animal Growth and Meat Technology. https://doi.org/10.1016/B978-0-12-815277-5.00010-X

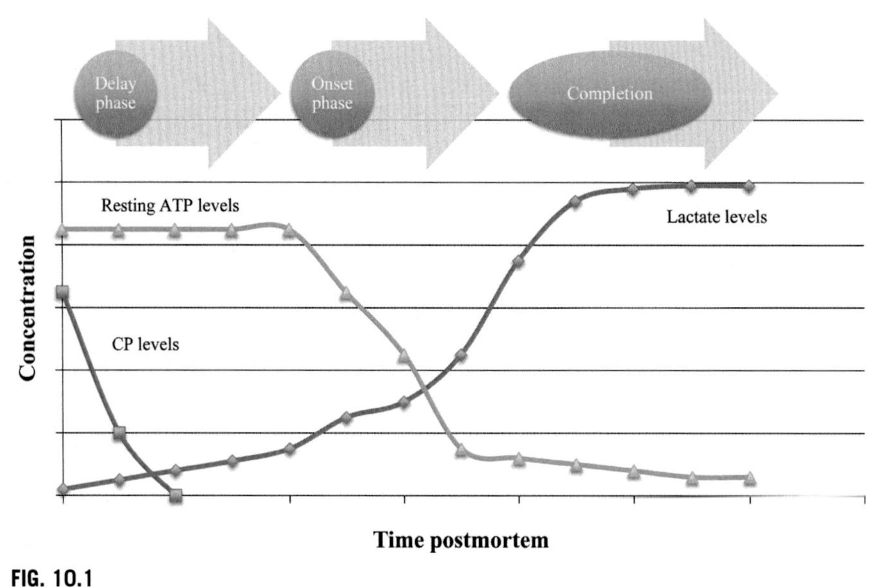

FIG. 10.1

Phases of the onset of rigor mortis.

can remain constant, the muscle will stay in the delay phase, with few noticeable changes. As muscle progresses through the onset of rigor and then finally completion of rigor, more actomyosin cross-bridges will form, the sarcomere will shorten, and there will be a noticeable loss in extensibility of muscle.

ATP has many functions, but its two primary functions in muscle include muscle contraction and calcium homeostasis. Hydrolysis of the high-energy terminal phosphate bond of ATP provides energy for muscle contraction. The energy released at the head of the myosin molecule allows the interaction of myosin and actin to form actomyosin cross-bridge attachments. These cross-bridge attachments must be broken (detached) to allow the continuation or cessation of muscle contraction. Energy from ATP is also used by specialized pumps to remove Ca^{2+} from the sarcoplasm (and myofibrils), which removes the calcium stimulus for contraction. Therefore ATP is required to prevent actomyosin cross-bridge formation and is required for the dissociation of actomyosin cross-bridges. ATP prevents rigor mortis. As ATP becomes depleted in postmortem muscle, both of these functions are lost, calcium is released across the sarcoplasmic reticulum membrane (and not pumped back into the sarcoplasmic reticulum by ATP-dependent pumps), and there is an accumulation of irreversible rigor bonds in sarcomeres across every muscle fiber.

In the first stage of rigor mortis, termed the delay phase of rigor mortis, postmortem muscle is in a soft, pliable, elastic, and extensible condition. This condition is referred to as prerigor. ATP (complexed with Mg^{2+}) levels are reasonably constant as the high-energy phosphate from creatine phosphate is expended (Fig. 10.1). Then there is a second phase of rigor mortis, the onset phase, involving a brief time

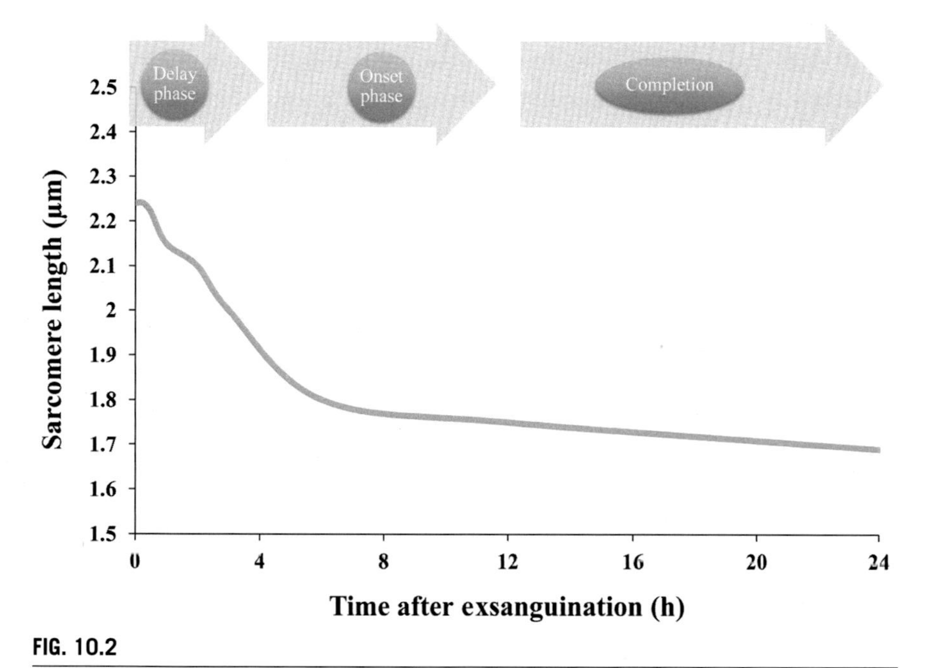

FIG. 10.2

Sarcomere length during delay, onset, and completion of rigor in lamb longissimus muscle.

From Wheeler and Koohmaraie, 1994, Journal of Animal Science, vol. 72, p. 1232.

of accelerated physical changes similar to those noticed in the first phase and proceeding to the point of the muscle being inelastic and inextensible. This is termed "in rigor." Measurable ATP depletion, lactate production, and sarcomere shortening (Fig. 10.2) occur during this phase. The third and final phase of rigor mortis (completion) occurs when postmortem muscle ATP levels are depleted and the actomyosin cross-bridges are thus irreversible. Generally, there is some rigor shortening at this point, and this is when meat is the least tender. Over time, meat does become more tender with postmortem aging. This is due in part to some resolution of rigor (and release of some actomyosin cross-bridges) but mostly due to protein degradation (Chapter 9).

CHEMICAL CHANGES IN POSTMORTEM MUSCLE

The glycolytic pathway is essential in the physiology of living muscle. In living tissues, including muscle, the glycolytic pathway converts glucose (primarily from glycogen) to two molecules of pyruvate. In this process, three molecules of ATP are generated. The product of glycogenolysis by glycogen phosphorylase is glucose-1-phosphate. This is an early intermediate in the glycolytic pathway and is the critical step in converting glycogen energy to ATP. In living muscle, pyruvate is metabolized by oxidative metabolism (Krebs cycle and oxidative phosphorylation) to generate more ATP. The 2 pyruvate molecules generated from the one molecule of glucose are oxidized to create 34 molecules of ATP. This points out the importance of oxidative metabolism in capturing the energy present within glucose and glycogen.

Accumulation of pyruvate in the cell is not favorable. In the aerobic condition, pyruvate is converted to acetyl-CoA by the pyruvate dehydrogenase complex. In the absence of oxygen, pyruvate still must be cleared. This is accomplished by lactate dehydrogenase to produce lactate. This is why lactate is the primary product of glycolysis in anaerobic postmortem muscle.

The progression of postmortem glycolysis can be tracked by evaluating the concentration of different metabolites in muscle. Changes in metabolites for living muscle and postrigor muscle (24 h postmortem) are presented in Table 10.1. Dramatic change occurs in muscle metabolites during this time frame, especially the decreased concentrations of glycogen, glucose, and ATP and the increased concentration of lactate and inosine monophosphate (inosine monophosphate is the product of deaminated AMP). Of course, the most important of these is lactate, and this is the explanation for a drop in pH in postmortem muscle from a living muscle pH of approximately 7.3 to approximately 5.6. The normal pH decline for each species can be generally characterized by a gradual decline in pH (Fig. 10.3; curve A). It is generally held that achievement of ultimate pH (regardless of the extent of pH decline) indicates the cessation of glycolysis and production of ATP and the onset of rigor mortis. The timing of onset of rigor mortis can vary across species (Fig. 10.4).

Other chemical changes have already been discussed in the context of physical changes. ATP declines (Tables 10.1 and 10.2) within a matter of hours because there is not sufficient oxygen in the muscle when the blood is removed to stimulate metabolic pathways to regenerate ATP in postmortem muscle. A certain level of ATP can be maintained shortly after death, however, because of its resynthesis from adenosine diphosphate and creatine phosphate reserves (Fig. 10.1; Table 10.1). These sources for ATP resynthesis, however, are quickly exhausted, and ATP content falls far below

Table 10.1 Changes in postmortem muscle metabolites in beef longissimus muscle

Metabolite	Concentration in μmoles/g fresh tissue	
	Living muscle	After rigor
Glycogen (obtained by acid hydrolysis)	56.7	1.4
Glucose	7.9	17.9
Lactic acid	13.9	84.6
Total acid soluble phosphorus (mainly nucleotides and creatine phosphate)	54.9	54.2
ATP	6.4	<1.0
Creatine phosphate	9.5	<2.0

From Bodwell, C. E., A. M. Pearson, J. Wismer-Pedersen, and L. J. Bratzler, 1965, Postmortem changes in muscle I. Chemical changes in beef muscle, Journal of Food Science, vol. 30, p. 766.

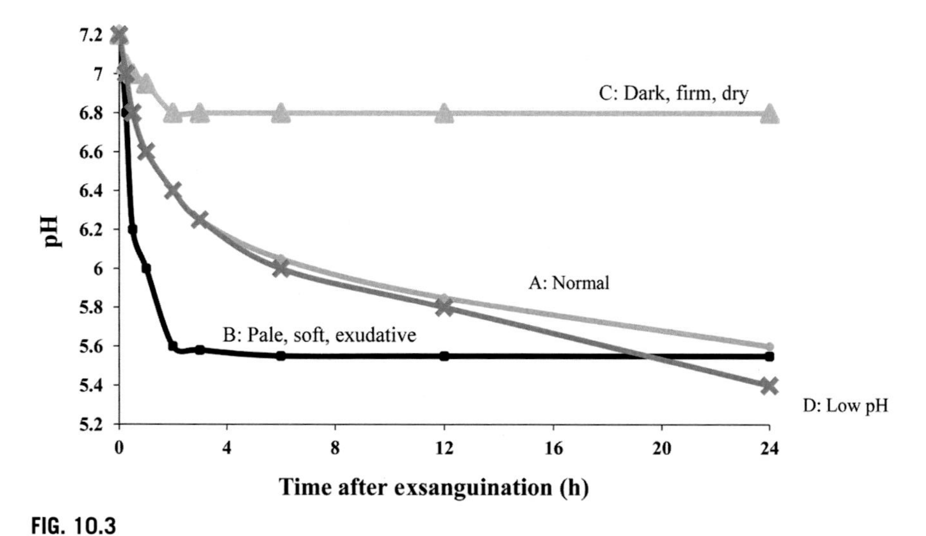

FIG. 10.3

pH decline in postmortem pork muscle.

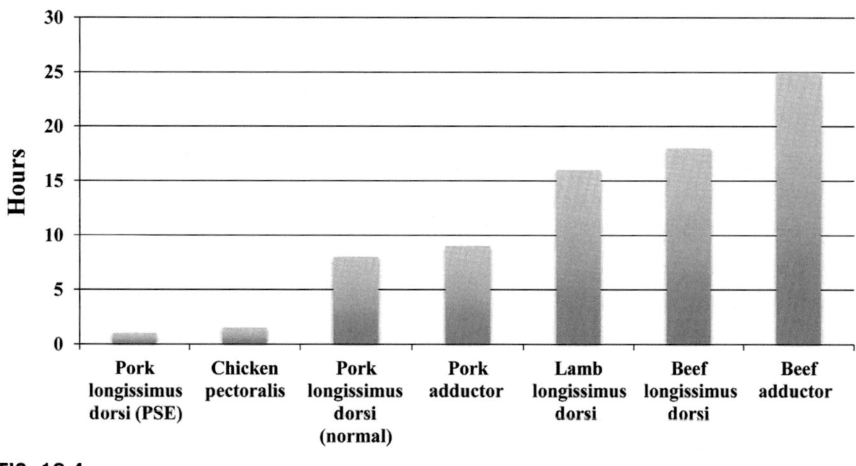

FIG. 10.4

Typical onset of rigor (defined by hours to reach pH 5.5–5.7) in various species and muscles.

From Honikel, 2004, "Glycolysis" in Encyclopedia of Meat Sciences, Elsevier Academic Press.

physiological levels (Table 10.1), climaxing its role in contraction and relaxation. Therefore in postmortem muscles there is an irreversible formation of rigor links in the actomyosin cross-bridges because of ATP loss. When this occurs, postmortem muscle is then inextensible and exists in a form of an irreversible muscle contraction. As stated earlier, this is the rigor condition. It is important to recognize this connection between the chemical condition and physical condition of postmortem muscle.

Table 10.2 Nucleotide levels in muscles with fast and slow rates of postmortem glycolysis

Item	Slow pH decline	Rapid pH decline
ATP (μmol/g)		
0-min postmortem	4.28	2.54
60-min postmortem	2.66	0.02[a]
ADP (μmol/g)		
0-min postmortem	1.54	1.60
60-min postmortem	1.58	1.17
IMP (μmol/g)		
0-min postmortem	0.54	0.95
60-min postmortem	1.23	2.46[a]

[a] *Significantly different within a row.*
From Kastenschmidt et al., 1966, Nature, vol. 212, p. 288.

VARIATION IN POSTMORTEM METABOLISM

RAPID pH DECLINE

The rate and extent of pH decline in early postmortem muscle can vary, and this can have profound effects on meat quality. Rapid lactate production (Fig. 10.5) and decline in pH (Fig. 10.3) in postmortem muscle can create part of the condition that is very damaging to meat quality. In fact, this is really only half of the equation. With rapid pH decline, there is a situation created that includes warm temperatures (still over 100°F) and acidic conditions. These conditions damage proteins that contribute to meat color (myoglobin) and water-holding capacity and texture (many myofibrillar proteins). The result is meat that has poor color (pale), poor texture (soft), and poor water-holding capacity (exudative). This condition is generally termed PSE (pale, soft, and exudative) and can occur in pork and in poultry meat.

There are three general causes of this rapid pH decline in pork and poultry: (1) genetic predisposition to stress, (2) acute stress immediately antemortem, and (3) slow chilling of carcasses during early postmortem processing.

There are well-documented mutations in a calcium channel protein (the ryanodine receptor) in the sarcoplasmic reticulum in pigs and turkeys. These mutations change the function of the protein and allow calcium to leak out of the sarcoplasmic reticulum and into the sarcoplasm. Remember that calcium will stimulate contraction but also is a message to activate glycolysis in several ways but primarily by stimulating glycogen phosphorylase. This condition in early postmortem muscle results in a rapid rate of glycolysis. Rapid glycogen breakdown means rapid lactic acid production (Fig. 10.5), depletion of ATP (Table 10.3), and, of course, pH decline (Fig. 10.3, curve B).

Rapid pH decline in turkey breast meat can cause denaturation of proteins that bind water, resulting in a loose-structured product that has poor water-holding capacity (Table 10.3). In extreme cases, pH can be as low as 5.5 within 30-min postmortem.

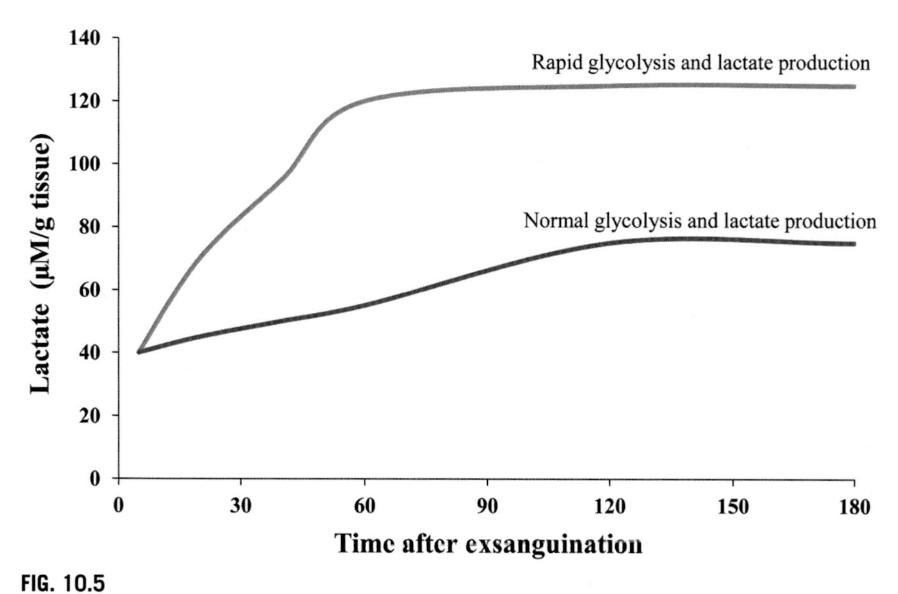

FIG. 10.5

Lactate production in postmortem muscle with slow and fast rates of glycolysis.

From Kastenschmidt et al., 1966, Nature, vol. 212, p. 288.

Table 10.3 Differences in traits in turkey breast muscles that have different rate and extent of pH decline

Item	High pH	Low pH
20-min postmortem		
pH	6.44	5.88
ATP (µM/g)	3.3	1.4
Lactate (µM/g)	58	94
180-min postmortem		
pH	5.88	5.69
ATP (µM/g)	1.6	0.4
Meat quality traits (24-h postmortem)		
Water-holding capacity (% yield; greater indicates better water binding)	112	85
Cook yield (%; greater indicates better yield and better bind by the meat proteins)	126	107
Minolta L^* (greater value indicates lighter color)	44	49

From Barbut et al., 2008, Meat Science, vol. 79, p. 46–63.

Of course, the onset of rigor occurs more rapidly in these conditions as well. It has been understood for some time that these are genetic conditions. With regard to pigs, the primary mutation in the ryanodine receptor has been eliminated from commercial breeding herds using genetic selection techniques. This was done first by determining genetic condition by testing the response of pigs to halothane. This is why the stress-susceptible condition was once called halothane positive. There are several noted mutations in the ryanodine receptor in turkeys.

Even in the absence of genetic predisposition for rapid pH decline, environmental factors (specifically acute stress immediately before slaughter) can result in a rapid pH decline. The acute stress can activate release of epinephrine. This endocrine message will stimulate the activation of glycolytic enzymes (specifically glycogen phosphorylase) to rapidly convert glycogen to ATP to provide fuel to address this stressful condition. When these enzymes are activated before slaughter, the rate of glycolysis is greater in postmortem muscle, and thus the rate of pH decline is greater. Although this usually does not result in the extreme pH decline as in the case of the genetic conditions discussed earlier, it is clear that acute stress before slaughter can result in poor color and water-holding capacity. For example, use of an electric prod can stimulate this type of acute stress that will decrease early postmortem pH in muscle, decrease pork water-holding capacity, and result in a pale color. Fig. 10.6 illustrates this point. In this experiment, hot shot use before slaughter resulted in lower pH at 0- and 45-min postmortem but not at 24h. This means that the total glycogen was not different at slaughter. Thus the extent of pH decline was not different, but the rate of pH decline certainly was. This more rapid pH decline resulted in pork that had greater drip loss (poorer water-holding capacity; Fig. 10.7) and greater L^* values (lighter in color; Fig. 10.8). If one had only measured the ultimate pH, they would not have been able to identify the reason (rapid pH decline) for the poor quality that resulted from the hot shot use.

As pointed out earlier, the reason that rapid pH decline is a concern for pork and poultry quality is that that combination of acid conditions and high temperatures

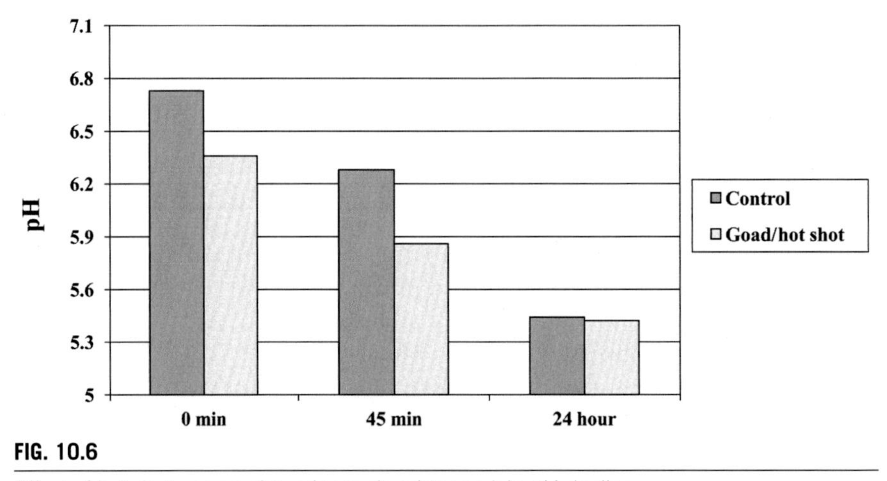

FIG. 10.6

Effect of hot shot use on pigs prior to slaughter on loin pH decline.

From Küchenmeister et al., 2005, Meat Science, vol. 71, p. 690.

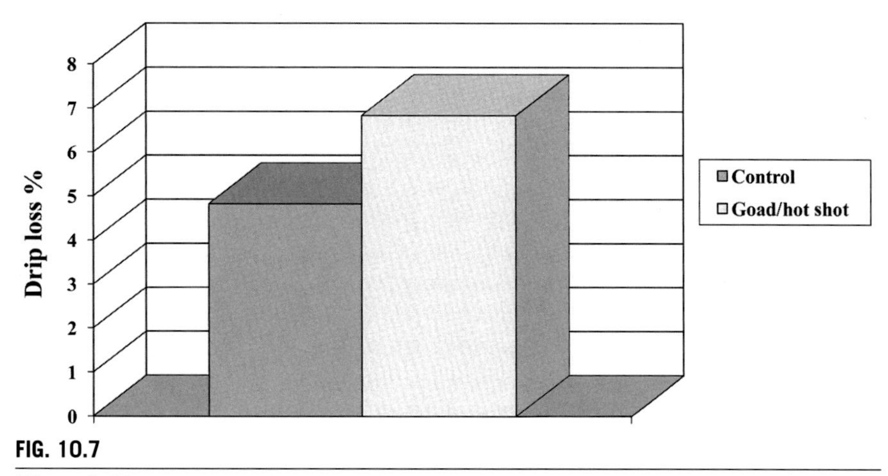

FIG. 10.7

Effect of different stress before slaughter on drip loss.

From Küchenmeister et al., 2005, Meat Science, vol. 71, p. 690.

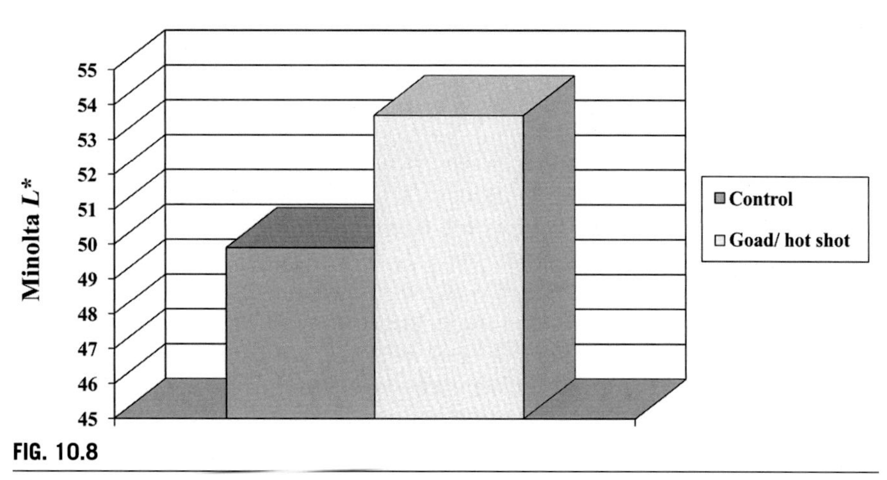

FIG. 10.8

Effect of different stress before slaughter on color.

From Küchenmeister et al., 2005, Meat Science, vol. 71, p. 690.

denatures proteins. Therefore poor chilling can result in a combination of low pH (<5.6) while the temperature remains warm. In modern commercial facilities, rapid chilling of carcasses can help to avoid the damaging combination of high temperature with low pH. Furthermore, the progression of glycogenolysis and glycolysis is slower at lower temperatures. Glycogen debranching enzyme is very temperature sensitive, and the rate of activity is slowed considerably at 40°F. Therefore more rapid chilling of pork and poultry meat should decrease the rate of pH decline and the incidence of PSE meat.

HIGH ULTIMATE PH

Chronic preslaughter stress can deplete muscle glycogen stores in muscle at slaughter. With little energy (glycogen) to convert to lactate, the result is a rapid onset of rigor, a shallow pH decline curve, and a relatively high ultimate pH (above pH6.5; Fig. 10.3, curve C). This condition has a marked effect on the quality of meat. The color is *dark*, the texture is *firm*, and the consistency is *dry*. Pork meat showing these defects is thus called DFD pork. Beef carcasses exhibiting these traits are called "dark cutters." The meat appears dark for several reasons. The first is that very little myoglobin is denatured at this pH. The second and maybe more important reason is that at this high pH, the myofibrils are very good at binding water (see Chapter 9) and are still swollen with water. This physical change causes the meat to absorb more light, and therefore it appears darker. The uncooked meat is called firm because very little of the myofibrillar protein is denatured and the water-holding capacity is very good. Finally, the fresh product appears to be dry because the proteins and myofibrils are very good at binding water because of the high ultimate pH (Chapter 9). There are some negative features of this product. Of course, the dark color is unattractive to consumers. The high pH may result in meat that is more prone to more rapid growth of spoilage bacteria. The tenderness of dark cutting beef is inconsistent. Dark cutting beef receives a discount in assigned USDA quality grade. Most surveys report that dark cutting beef occurs in about 3% of beef in the United States. Pork or poultry that are considered DFD are very infrequently observed in commercial settings.

LOW ULTIMATE pH IN PORK

There are observations of pork that has an extended pH decline but no large difference in the rate of pH decline (Fig. 10.3, curve D). It is intuitive that a greater production of lactic acid must be due to a greater concentration of glycogen in muscle. This is caused by a mutation in the AMP-dependent protein kinase (AMPK). This protein acts as the fuel gauge for the cell but is also the regulator. The normal AMPK will inhibit glycogen synthesis when the energy status is adequate. It makes sense that AMP:ATP ratios will increase when the cells or tissues are stressed. The increase in AMP will activate AMPK and stimulate the energy storage pathways including glycogen synthesis. When AMPK is not activated, glycogen synthesis will be inhibited. However, the mutation in the muscle-specific regulatory subunit of the protein (γ^3 regulatory subunit, PRKAG3 gene) eliminates the ability of the AMPK to turn off glycogen synthesis. The result can be 70% more glycogen in the muscle (Table 10.4). In postmortem muscle, this means the ultimate pH may be very low (5.55 or less) and water-holding capacity will be very poor (Chapter 9).

The phenotype associated with this mutation was first noticed in Hampshire pigs, and for some time it was called the "Hampshire Effect." This has also been called the RN gene because the yield of Napole hams (Rendement Napole) was low when hams had this phenotype. This is where the nomenclature for this mutation comes from. There are two alleles: RN^- (the dominant allele associated with greater glycogen content and low pH) and rn^+ (the wild-type recessive allele). Some characteristics of pork from pigs with these genotypes are included in Table 10.4. In general, low pH and poor cooking yield are observed. These are typically considered as poor quality traits. However, these traits might be considered to be useful in production of dry cured hams.

Table 10.4 Effect of RN genotype on composition and quality of the longissimus muscle of pigs

Item	rn^+/rn^+	RN^-/RN^-	RN^-/rn^+
Glycolytic potential at slaughter (μmol/g)[a]	108[b]	222	195
Ultimate pH in loin[a]	5.74[b]	5.54	5.53
Pork loin L score (0–100; 0 = black, 100 = white)[a]	47.8[b]	51.2	50.3
Drip loss %[c]	5.17[b]		6.71
Cook loss %[c]	29.64[b]		34.48
Napole yield % (ham processing yield)	63.73[b]		59.53

[a] *Lebret et al., 1999, Journal of Animal Science, vol. 77, p. 1482.*
[b] *Means for the wild-type genotype (rn^+/rn^+) were significantly different from the RN^-/RN^- and RN^-/rn^+ genotypes.*
[c] *Bertram et al., 2003, Meat Science, vol. 65, p. 707.*

COLD SHORTENING

Although ATP is high and in the presence of Ca^{2+}, unrestrained prerigor muscle held at 35°F will dramatically shorten—sarcomeres are super contracted. This is called cold shortening. Cold conditions in muscle (<45°F) before rigor create this condition. The reason is that as the temperature declines, the sarcoplasmic reticulum can no longer sequester Ca^{2+} from the myofibrils. Under extreme cases, Ca^{2+} may also be released from the mitochondria. The reason for this calcium dumping is that at cool temperatures the mostly lipid membranes of the organelles become brittle and cannot hold the calcium against the gradient that favors flowing into the sarcoplasm. Of course, this will ultimately happen during normal chilling of carcasses. The critical difference is that when this happens rapidly, ATP is still present. With plenty of calcium and ATP available, the myofibrils have all they need to shorten. This is more likely to occur in beef and lamb because both have relatively long delays to the onset of rigor (Fig. 10.4). In other words, there is plentiful ATP even when chilling has occurred. Meat from cold-shortened muscle will be very tough because of the high density of actomyosin (cross-bridges of myosin and actin formation) within the shortened sarcomeres of the myofibril, and cross-bridges cannot dissociate because there is no ATP available in the muscle.

The application of electrical stimulation technology (Table 7.3) immediately after slaughter of cattle prevents cold shortening of beef muscle and initiates tenderization processes by depleting ATP before the onset of rigor mortis. Because this also stimulates the glycolytic process and more rapid production of lactic acid, this procedure should not be used in pork or poultry.

THAW RIGOR

Another interesting phenomenon associated with prerigor muscle held in cold temperatures is thaw rigor. This occurs when prerigor muscle is excised, frozen, and subsequently thawed. During thawing the muscle shortens because there is enough residual ATP and Ca^{2+} to initiate shortening. As a result, excessive quantities of fluids are squeezed from the muscle during shortening. From a practical standpoint to avoid fluid loss, lightweight carcasses and cuts should not be frozen until rigor is resolved (after 24-h postmortem) under normal aging conditions.

SUMMARY

The physical and chemical changes in the conversion of muscle to meat have a dramatic effect upon the quality characteristics of meat. Indeed, the fate of these events starts with the physiological and biochemical state of the living animal. Along with the biology of meat (skeletal muscle), animal production, management, environment, genetics, preharvest handling, and postmortem storage and handling are major determinants of changes in meat and its ultimate market value. From a postmortem conversion of muscle to meat perspective, the rate and extent of postmortem glycolysis may be the most important factor affecting the meat quality characteristics of color, water-holding capacity, tenderness, and economic return. Also, postmortem proteolysis of certain structural proteins of skeletal muscle plays an important role in tenderness, especially beef tenderness. These physical and chemical changes taken together ultimately modify market value and consumer demand and satisfaction, and these changes must be controlled and monitored by using proper production, processing, and merchandizing techniques to attain high-quality meat and meat products.

QUESTIONS FOR STUDY AND DISCUSSION TOPICS

1. What is the role of ATP in maintenance of muscle in the relaxed state?
2. What are two roles of calcium in the conversion of muscle to meat?
3. Draw a pH decline curve for the following situations:
   a. Chronic stress before slaughter,
   b. Acute stress before slaughter,
   c. Use of electrical stimulation in a beef carcass early postmortem, and
   d. In muscle of pig that is a RN^{-}/RN^{-} genotype.

CHAPTER

Muscle structure and function

11

INTRODUCTION

Learning about the structure and function of muscle, especially skeletal muscle, is essential for understanding the practical significance of its food product, meat. Furthermore, muscles are a highly sophisticated system in the living animal designed for movement (skeletal), digestion (smooth), and respiration (heart). It is also important to recognize the role of skeletal muscle after its conversion to meat and subsequent entrance into domestic and global markets.

SKELETAL, SMOOTH, AND CARDIAC MUSCLE

Three classes of muscle can be distinguished functionally and anatomically in the living animal: skeletal, smooth, and cardiac muscles. Skeletal muscle is defined as muscles attached to the skeleton. It is the most abundant of the three kinds, making up about 40% of the bodyweight of animals. Skeletal muscle is responsible for the movement system of the body and functions under voluntary control by the nervous system. Skeletal muscle cells are multinucleated (100–200 nuclei per cell). Smooth muscle cells are mononucleated. Smooth muscle is found primarily in the walls of the digestive tract, viscera, reproductive tract, and blood vessels and is under involuntary nervous control. Cardiac muscle is found only in the heart and is controlled by the involuntary nervous system. When observed microscopically, another distinguishing characteristic of skeletal and heart muscle is the regular array of striations, whereas smooth muscle is nonstriated. Cardiac muscle cells usually contain only one nucleus and are separated by intercalated disks that aid in the synchronized contraction of regions of the heart.

SKELETAL MUSCLE CHARACTERISTICS

Muscles are classified as organs and as such have functional components like other organs. Fig. 11.1 depicts the organizational structure of a skeletal muscle.

The diagram shows a cross section of muscle enveloped by the connective tissue layer called the epimysium. The figure also includes the muscle organization of (1) muscle fascicles (bundles) surrounded by the connective tissue layer called the perimysium and (2) the muscle fibers (individual muscle cells) surrounded by the connective tissue layer called the endomysium, shown in greater detail in Fig. 11.2.

The Science of Animal Growth and Meat Technology. https://doi.org/10.1016/B978-0-12-815277-5.00011-1

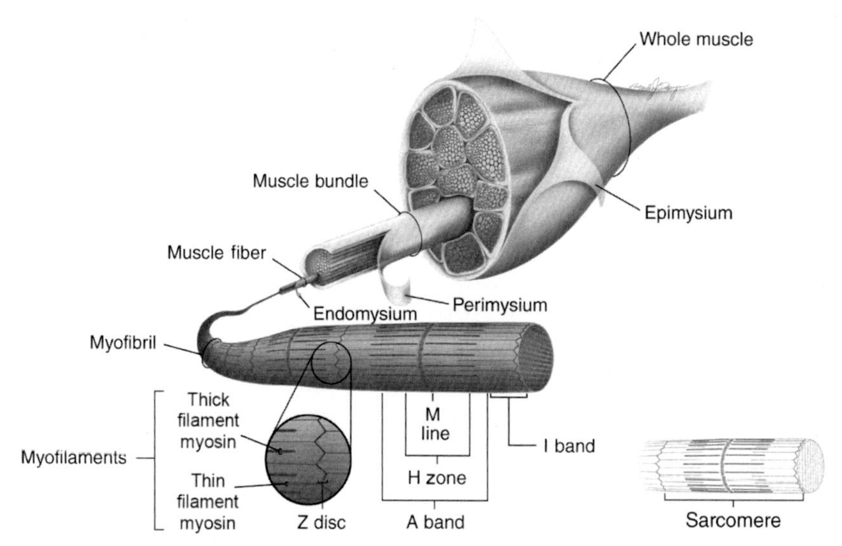

FIG. 11.1

A schematic representation of a skeletal muscle in cross section showing the organizational structure of the muscle bundles (fasciculi) and fibers with their respective surrounding layers of connective tissue.

Diagram by Marley Dobyns, Animal Science Department, Iowa State University.

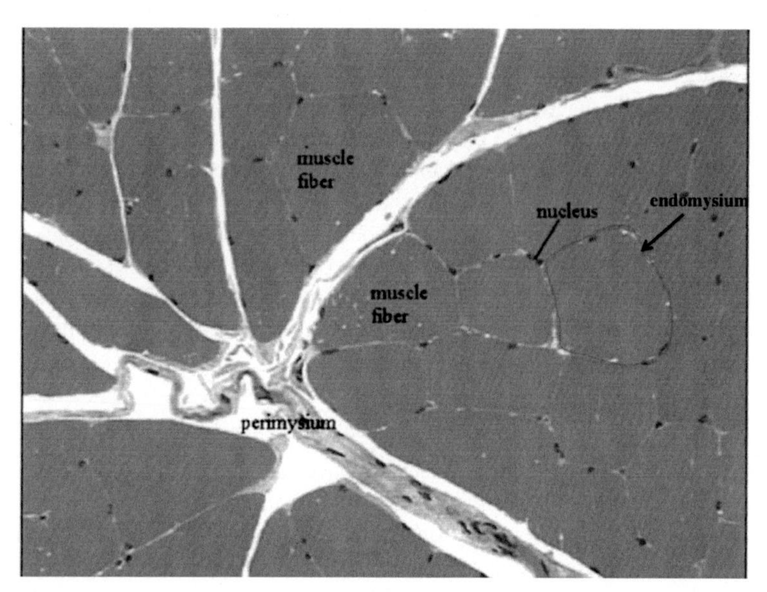

FIG. 11.2

A photomicrograph of a cross section of skeletal muscle showing muscle fibers *(circled)* within a muscle bundle surrounded by perimysial connective tissue.

Picture courtesy of the Muscle Biology Group, Department of Animal Science, Iowa State University.

The connective tissue layers are thicker in muscles used for power compared with muscles used for coordination. Also, these thicker layers (sheaths) of connective tissue found in power muscles contribute to less tender meat.

Branching off and continuous with the epimysium is a layer of connective tissue surrounding the muscle bundles (fasciculi) called the perimysium. An example is shown in Fig. 11.1. There are both primary and secondary bundles, and the larger the fasciculi, the coarser the texture of the muscle. Power muscles performing large movements (legs) have larger bundles (coarser texture) relative to smaller muscles performing fine movement (back). The perimysium also contains and envelops blood vessels and nerves. The structured location of another layer of connective tissue, termed the endomysium, is shown in Fig. 11.1 and branches from the perimysium. This is the thin layer of connective tissue that loosely surrounds muscle fibers (cells), the physiological unit of skeletal muscles. Therefore from the myotendinal junction, collagenous fibers from the tendon connect directly to the muscle cell membrane. The endomysium is adjacent to the sarcolemma, the cell membrane. The sarcolemma is the muscle cell membrane responsible for the transfer of chemicals and conduction of the electrical impulse necessary for contraction. Also, you can observe in the cross section of muscle other constituents such as blood vessels, capillaries, and inter- and intramuscular fat.

MUSCLE FIBERS

Muscle fibers (cells) are the basic unit of muscle. They are long, cylindrical, tubular cells with tapering conical ends, unbranching, and not perfectly round in cross section. The number of fibers varies from 50 to 300 in each fasciculus (bundle). Power muscles have larger fasciculi and larger fibers but fewer fibers per bundle. The length of muscle cells can range from 1 mm to 34 cm (a centimeter is 0.4 of an inch and a millimeter is 0.1 cm), although long cells are rare; most of them range from 1 to 40 mm with an average of 20 to 30 mm. The diameter of muscle cells can vary from 10 to 100 μm (μm), or microns, because cells taper at each end. Fiber diameter is affected by many factors. For example, fiber diameters are larger in fish than in mammals, and fibers in mammals are larger than those in birds. Shorter, thicker muscles have fibers that are larger in diameter than longer, thinner muscles. Exercise, increased maturity, and higher plane of nutrition all contribute to the fiber profile. Increased aerobic activity will increase capillary density, mitochondria number, and even myoglobin content. Increased anaerobic activity (sprinting, lifting weight) will increase fiber diameter and increase the proportion of fast-twitch fibers. Of course, genetics can influence this development. It is easy to see that cattle breeds that were once used for draft have a larger frame and often a differing profile of fibers. This is because of numerous generations of selection for strength and endurance.

Skeletal muscle is made up of different fiber types dependent on their visual appearance, metabolism, speed, or strength of contraction. From a gross morphological classification, fibers can be characterized by their color (red or white) in the muscle (see Fig. 9.6). Red muscle fibers, found in the endurance muscles of the leg and shoulder, appear red because they have great concentrations of myoglobin (the pigment in muscle and meat). They are further characterized by their long, slow, sustained contractile activity and by their high oxidative (aerobic) enzyme activity. Thus red muscle fibers have a greater concentration of mitochondria associated with their oxidative activity. Conversely, those fibers with low concentrations of myoglobin found in muscles such as porcine longissimus (loin) or poultry pectoralis major (breast) muscles are termed white muscle fibers. These fibers contract quickly and for a short duration. For this type of contractile activity, white fibers are dependent upon glycolytic metabolism (anaerobic), large stores of glycogen (noted sometimes as glycogen granules), rapid rates of glycolysis, and a highly developed sarcoplasmic reticulum as depicted in Fig. 11.3. The sarcoplasmic reticulum is a network of membranous tubules that have the ability to bind and release Ca^{2+} and are involved in the mechanism of muscle contraction. Intermediate fibers have properties in between those of red and white fibers, hence the name intermediate. Indeed, few muscles contain muscle cells of just one fiber type.

Based on speed of contraction as a classification, slower contracting fibers are classified as Type I, whereas faster contracting fibers are characterized as Type II.

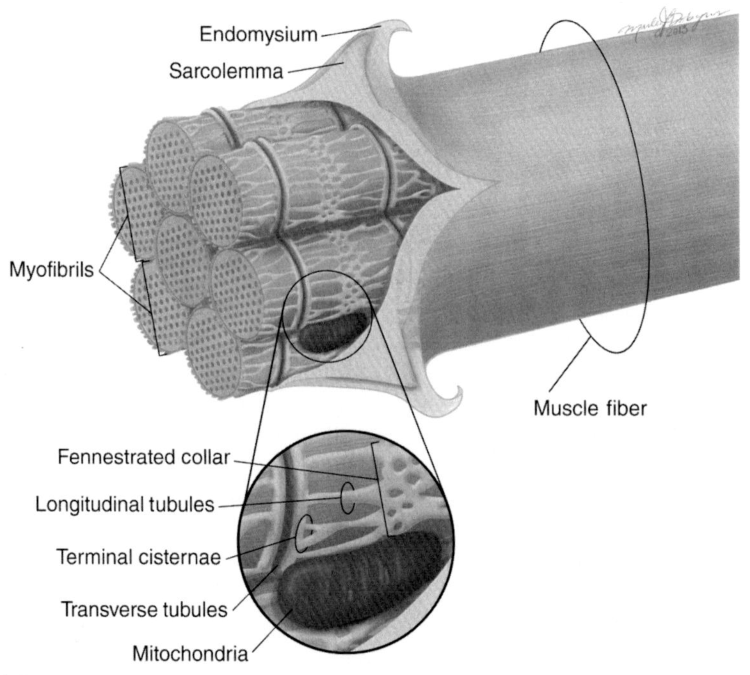

FIG. 11.3

A diagrammatic representation of the sarcoplasmic reticulum within a skeletal muscle fiber. The cell membrane has been peeled back to show the sarcoplasmic reticulum.

Diagram by Marley Dobyns, Animal Science Department, Iowa State University.

Other classifications of fiber types are more complex and sophisticated. That is, classification can be based on their inherent glycolytic or oxidative metabolism, and myosin ATPase activity. With the use of histochemical analysis, fiber types react differently to certain oxidative, glycolytic, and ATPase enzyme-specific stains. By using the ATPase stain, three different fiber types can be observed: red, white, and intermediate as described in Table 4.1. Differences in these fiber types can then be observed, measured, and photographed by using an appropriately equipped light microscope.

MYOFIBRILS

Fig. 11.4 provides a schematic of the hierarchal relationship of whole muscle to a relaxed myofibril. Muscles are organs consisting of muscle fibers (cells) described in Figs. 11.1 and 11.2. Fibers or cells are specialized in structure and function in that their structure is highly organized and their function involves contractile activity.

When observed in the electron microscope, there is a regular array of alternating dark and light banding patterns (Fig. 11.5). Fibers are made up of myofibrils, and myofibrils (Fig. 11.4) are subcellular structures specialized in contractile activity. Myofibrils make up more than 50% of total protein in the cell and are not encased in a membrane. The proteins constituting the myofibrils are insoluble at the ionic strength of the cell, and this is the reason that myofibrils can be observed microscopically as structural entities. Adjacent myofibrils lie with their light and dark bands in register, and as a result confer a cross-striation appearance upon the entire cell. As observed in Fig. 11.6, a relaxed myofibril has repetitive bands. Under postmortem conditions, the relaxed myofibril stage results in more tender meat than the constricted stage. The dark band, or A band, is anisotropic (birefringent) and hence is dark in the phase (light) microscope but bright (birefringent) in the polarizing microscope. The light band, or I band, is isotropic and is light in the phase but dark (nonbirefringent) in the polarizing microscope.

The dark, thin, vertical line bisecting the I band is called the Z disk (line). The sarcomere is the structure from one Z line to the next within the myofibril (Fig. 11.4). And this distance in a relaxed sarcomere is 2.4–2.8 μm in muscle from almost all domestic species. When muscle contracts, sarcomere lengths decrease to 2 μm, or even less. The myofibril is a series of several hundred to thousands of sarcomeres within the muscle cell, depending on the length of the muscle cell.

A light area in the middle of the A band is called the H (heller) zone, observed only when the myofibril is in a relaxed state. Myofibrils are 1–3 μm in diameter and may occasionally branch. The M line in the center of the H zone, the pseudo H zone, and thick, thin, titin, and nebulin filaments can only be observed for details with the electron microscope. Table 11.1 provides information about the components making up the microscopic appearance of the myofibril. The abbreviations using letters for these components come from German words used by early microscopists, for example, Z line (disk) = zwischenscheibe.

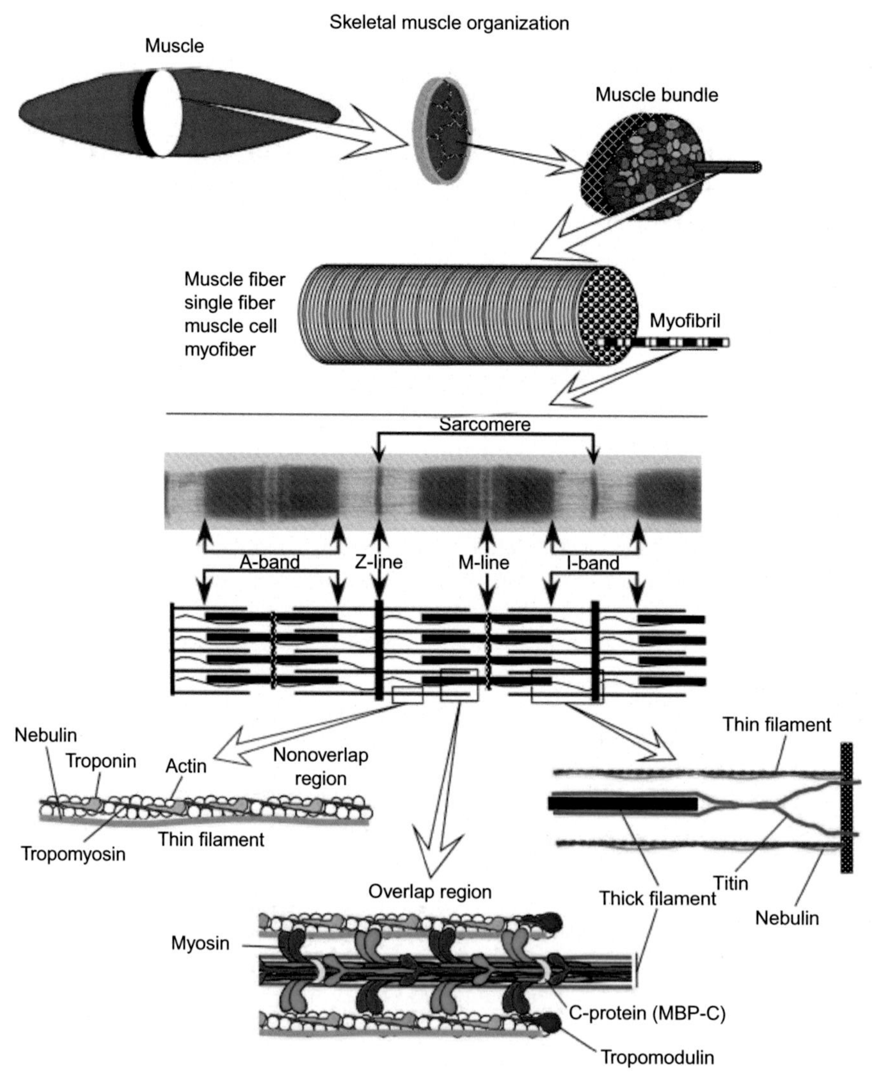

FIG. 11.4

A schematic representation of the organization of components of a skeletal muscle down to the contractile elements of a myofibril.

Diagram courtesy of the American Meat Science Association, drawing by Darl Ray.

SUMMARY

The purpose of this chapter is to show the importance and relationship of the structure and function of living and postmortem muscle. The knowledge, understanding, and application of muscle structure and function to postmortem muscle are fundamental in securing the quantitative and qualitative characteristics desired in meat and meat products. The structure of muscle is presented to show the gross anatomical

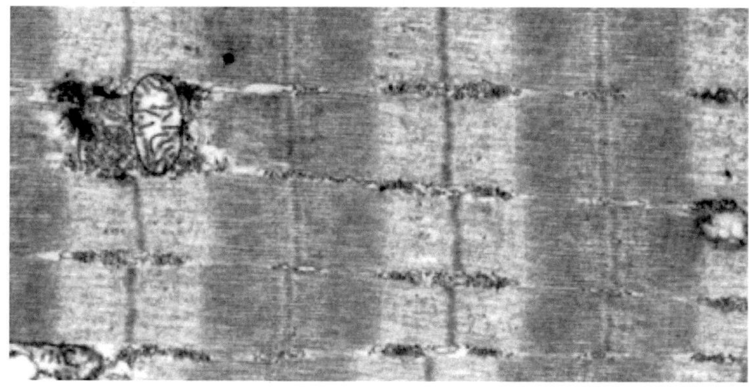

FIG. 11.5

An electron micrograph of a longitudinal section of a skeletal muscle myofibril showing the contractile proteins and distinctive Z lines. Note the intricate assembly and the arrangement of adjacent myofibrils that allows for coordinated contraction within fibers and, eventually, muscle bundles and muscles.

Photograph courtesy of the Muscle Biology Group, Department of Animal Science, Iowa State University.

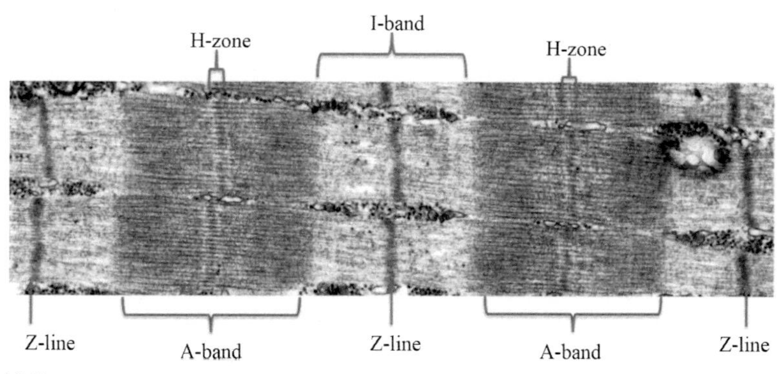

FIG. 11.6

An electron photomicrograph of a longitudinal section of a myofibril from skeletal muscle with specific areas and structures labeled.

Photograph courtesy of the Muscle Biology Group, Department of Animal Science, Iowa State University.

Table 11.1 Definition of terms used to describe the microscopic properties and appearance of striated skeletal muscle

Term	Definition
A band	Anisotropic, birefringent (bright) in polarizing, dark in phase
I band	Isotropic, nonbirefringent (dark) in polarizing, bright (light) in phase
Z line (disk)	Zwischenscheibe (between disks)
M line	Mittelscheibe (middle disk)
H zone	Heller (bright)

parts as well as the microscopic elements of muscle and their relationship to muscle contraction-relaxation and to postmortem meat quality characteristics, especially tenderness. The function of living muscle is that of contractile activity in the movement system. The remarkable discovery of thick and thin filaments and their interaction resulting in contraction-relaxation has proven to be valuable in meat science and the meat industry.

Postmortem skeletal muscle serves as a food, meat. Many paths and processes are taken to transform the carcass of an animal into wholesale and retail fresh and processed meat items for eventual consumer purchase. These items can take on the form of tender and less tender and cured and smoked retail cuts, ground meat, hot dogs, and jerky. The large number of different meat products is designed to address income level, ethnic background, and regional preferences and offers convenience, variety, texture, and taste.

QUESTIONS FOR STUDY AND DISCUSSION TOPICS

1. Name the three layers of connective tissue found in skeletal muscle.
2. Compare the structural similarities and differences of the three types of muscle.
3. What is the contractile unit of skeletal muscle and how long is this unit in relaxed muscle?
4. Describe the types of muscle fibers found in skeletal muscle and provide examples where each may be found.
5. Draw the structure of a myofibril in longitudinal section and label the unique areas as seen in the electron microscope.
6. Where are the mitochondria located within the muscle fiber?

CHAPTER

Meat microbiology and safety

12

INTRODUCTION

The consumption of safe, wholesome meat is a major issue with consumers. Along with safety, consumers want to purchase eye-appealing meat cuts in attractive surroundings. They also need to know how to keep meat safe and wholesome when they bring the meat purchases home. Unfortunately, consuming contaminated meat accounts for approximately 50% of the outbreaks of food-borne illnesses in the United States. Furthermore, food-borne illnesses are costly not only to the consumer (e.g., time lost at work, medical problems and insurance fees, etc.) but also to the meat processor in lost sales, legal fees, and settlement costs. Some food-borne illness cases are so severe there is a loss of life, and the media attention given to these cases may result in an overall downturn in meat consumption, at least for a short period of time. Moreover, financial loss from spoiled meat and meat products is considerable. With all of the food-borne illness and spoilage loss, it behooves the entire livestock and meat industry to produce safe, wholesome meat and meat products. Also, consumers must know how to handle meat using good hygienic practices and cookery methods to prevent spoilage and illness and have delicious meat and meat products to enjoy.

There are three types of cleanliness and hygiene (physical, chemical, and microbiological) that must be diligently implemented to help ensure safe, wholesome meat and meat products. Physical methods may be the easiest form of cleanliness to perceive and carry out as this refers to, as examples, good dressing practices during slaughter and keeping meat cutting and processing tables clean. Chemically clean is defined, for example, as the removal of detergents and sanitizers from cleaned surfaces and equipment. Inadequate removal of chemicals from equipment or tables could result in costly product quality defects. Microbiologically clean denotes controlling, reducing, and/or eliminating bacteria, yeast, and molds under all conditions of meat production, processing, distribution, and consumption. Bacteria represent the largest group of the three microorganisms and are by far the one that causes the most food spoilage and illness.

The benefits of proper cleanliness are immense. With proper cleanliness, meat will be safer and have a longer shelf life and have improved sensory and organoleptic quality, decreased product loss, and enhanced public image. In the final analysis, the cumulative benefits of cleanliness mean a more profitable business. It pays to be "clean" when operating a meat business!

The Science of Animal Growth and Meat Technology. https://doi.org/10.1016/B978-0-12-815277-5.00012-3

MICROBIOLOGY

Microorganisms are often thought of as germs. Some people even call them "bugs," but they are not insects. What are they? Microbes are single-cell plants that cannot be seen by the naked eye, with the exception of molds. It then usually takes a microscope to observe and identify single-cell organisms, therefore the term microorganisms. Microorganisms are found everywhere—air, soil, water, farms, animals, feces, processing plants, retail stores, restaurants, homes, and people. Indeed, people are one of the main sources responsible for transmission and control of microbial growth and contamination. Interestingly, most microorganisms are harmless. But some cause spoilage, illness, disease, and death, while others are beneficial and are actually used in meat processing formulations, such as using a lactic acid starter culture in summer-sausage production.

There are three classes of microorganisms; bacteria, molds, and yeasts. Because bacteria are by far the most important in meat spoilage and food-borne illness, they will be emphasized in more detail in this chapter.

MOLDS AND YEASTS

A few brief points will be made about the nature and properties of molds and yeast. Molds are multicellular microorganisms that can be visible as a colony (Fig. 12.1). They are made up of multiple types of cells that require oxygen for growth. Their optimum pH range is 4.5–6.8, although they can grow over a wide range of pH values, and their minimum water activity (a_w) requirement for growth is below 0.7. Hence,

FIG. 12.1

Example of mold (fungi) colonies on meat samples.

Courtesy of Dr. James Dickson, Department of Animal Science, Iowa State University.

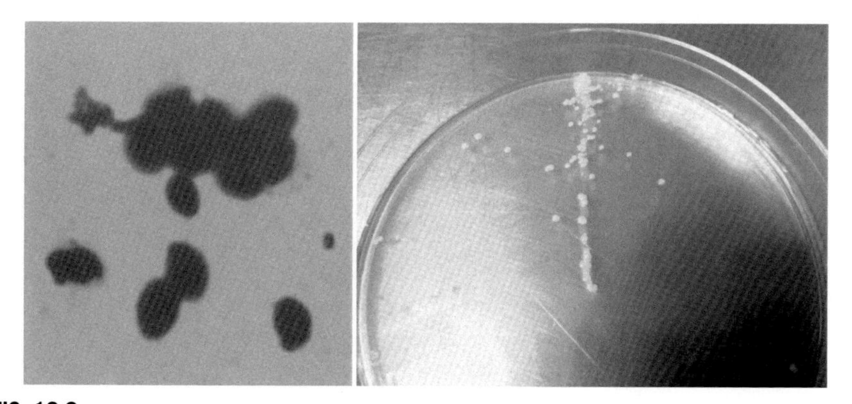

FIG. 12.2

Examples of yeast cells and colonies on tryptic soy agar plates.

Courtesy of Dr. James Dickson, Department of Animal Science, Iowa State University.

molds can grow under dry conditions, and it is not unusual to find them on products such as dry country cured hams and dried sausages.

Most yeasts occur normally as spherical, ellipsoidal, or elongated cells (Fig. 12.2). Also, most yeasts reproduce by vegetative bud formation. Their minimum water activity (a_w) is below 0.8, and they have a typical growth range of pH 3.0 to above 7.5. They can grow with or without oxygen. Yeasts are beneficial in fermentation of sugar to alcohol and the production of carbon dioxide during bread making. Conversely, large populations of yeasts on fresh meat can cause meat spoilage.

BACTERIA

Bacteria are single-cell (unicellular) microorganisms that can grow under many different environmental conditions and are found everywhere. Bacterial cells are observed microscopically as rods and spheres and measure about 0.5–1.5 μm (Fig. 12.3). They absorb nutrients from their environmental surroundings and excrete mostly wastes, but some of the waste are beneficial substances, including flavor components and lactic acid during fermentations. Bacteria reproduce by binary fission, and cell elongation, division, and rapid growth can occur in minutes. For example, after 1, 2, 4, 8 h, and so on, they can continue to grow, and the numbers double with time. Obviously, growth can occur rapidly. Fig. 12.4 is a typical bacterial growth curve. In the first phase, the lag phase, the bacterial growth is slow, but once they start their growth it is incredibly fast. For example, after 8 h, one cell can become 16 million cells under optimal growth conditions. Their growth curve in the stationary phase is flat usually because of the accumulation of inhibitory waste products from excessive growth.

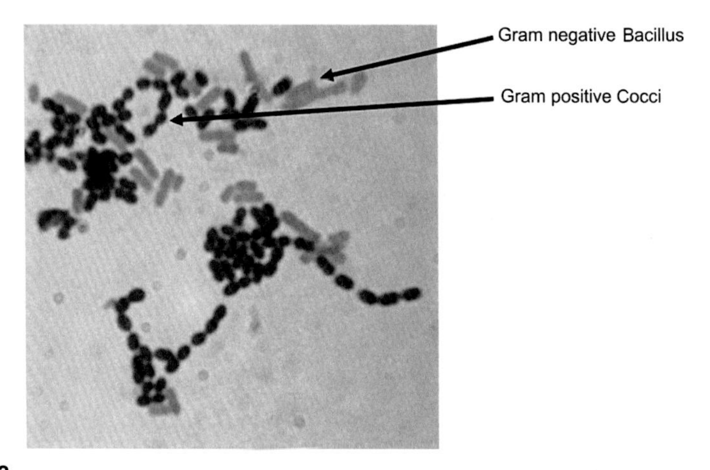

FIG. 12.3

Examples of gram-negative bacillus and gram-positive cocci bacteria.

Courtesy of Dr. James Dickson, Department of Animal Science, Iowa State University.

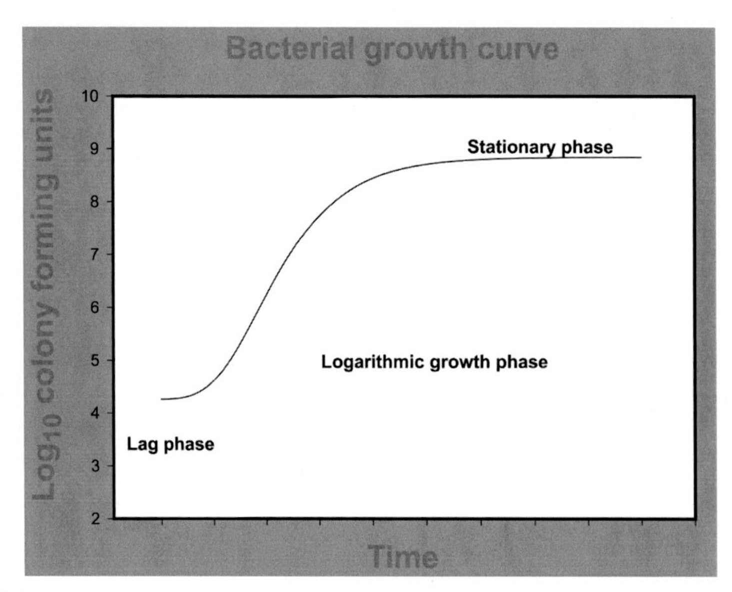

FIG. 12.4

Example of a typical bacterial curve.

Courtesy of Dr. James Dickson, Department of Animal Science, Iowa State University.

GRAM-NEGATIVE AND GRAM-POSITIVE BACTERIA

Bacteria can be identified as to their reaction with crystal violet stain, a common method of bacterial identification that differentiates between the basic cell wall structure of the bacteria. Bacteria staining red have cell walls that do not retain crystal violet and are called "gram negative," and those that do retain the dye stain purple are termed "gram positive." Hence, bacteria are grouped according to gram-negative or gram-positive reactions (see Fig. 12.3).

SPORE FORMERS

Some bacteria are characterized as spore formers because they produce spores when they are in an environment not conducive to normal growth and multiplication. In a nonconducive growth environment they produce spores, which are a dormant state of the bacterium. Then when conditions become favorable, spores germinate to form vegetative cells. These are then capable of growing and multiplying like other bacteria. *Clostridium botulinum*, the pathogen causing botulism, is an example of a spore former (Fig. 12.5). In the spore state, it can survive very cold or, in particular, very hot temperatures. Normally, these temperatures would destroy the vegetative state. The toxin of the spore former, when ingested, is the cause of botulism in humans.

AEROBES AND ANAEROBES

Bacteria have different oxygen requirements for their growth. They are characterized as aerobes (require oxygen), facultative anaerobes (grow with or without oxygen), microaerophilic (require 6% oxygen), and anaerobes (require no oxygen). Examples are shown in Fig. 12.6.

ACIDITY AND ALKALINITY

Acidity or alkalinity, measured by pH, is another factor affecting bacterial growth that can interact with oxygen. They prefer a neutral pH (pH 7.0) but generally have a range of 5.4–7.5 for growth in animal products. This wide range allows growth on

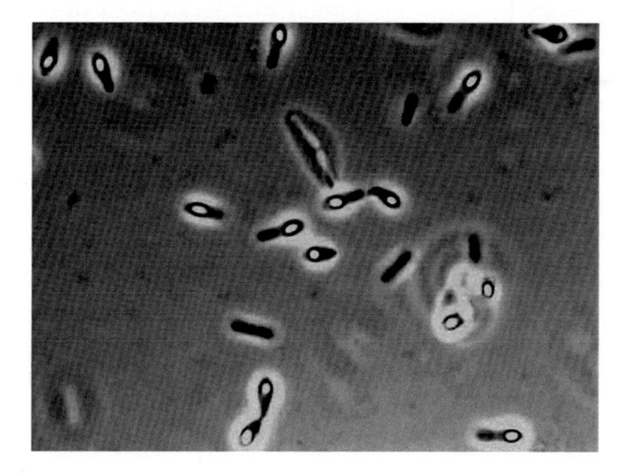

FIG. 12.5

An example of spore forming, *Clostridium botulinum*, bacteria.

Courtesy of Dr. James Dickson, Department of Animal Science, and Robert Hubert, Interdepartmental Program–Microbiology, Iowa State University.

Classification	Examples
Aerobic	*Micrococcus*, *Pseudomonas*, molds
Facultative anaerobic	*Escherichia*, *Enterobacter*, *Salmonella*, yeasts
Anaerobic	*Bacteroides*, *Clostridium*

Aerobic bacteria—require oxygen for growth

Facultative anaerobic bacteria—will grow with or without oxygen

Anaerobic bacteria—will only grow in the absence of oxygen

FIG. 12.6

Effect of oxygen levels on bacterial growth.
Courtesy of Dr. James Dickson, Department of Animal Science, Iowa State University.

many things including animals, facilities, meat products, and people and in the air, water, and soil. Fresh meat can have a normal pH range of 5.2–6.8. At pH 6.8, more bacteria will grow, but when the pH is low in meat products such as summer sausage (pH 4.6), almost no spoilage will occur.

PSYCHROPHILES

Another very important characteristic of bacteria is their growth temperatures. Those growing at refrigerated temperatures are classified as psychrophiles. These are true cold-loving microbes that grow at temperatures below 20°C (68°F). Commonly, these are the likely ones found growing on refrigerated meat. Psychrotrophs can grow at cold temperatures, but they grow best at mesophilic temperatures (20–50°C or 68–122°F). The importance of psychrophiles and psychrotrophs is that they are the major meat-spoilage bacteria. They can cause fresh meat discoloration and undesirable odors and impart off-flavors in cooked or fresh meat when growth is high. *Listeria monocytogenes* is a significant pathogen, especially in ready-to-eat foods, which can grow at refrigeration temperatures (Fig. 12.7).

MESOPHILES

Most bacteria, called mesophiles, grow from 20°C to 50°C (68°F to 122°F). Many pathogens are found in this group, and they are of major significance as they cause human disease and death. They grow best at room temperatures. Foods left at room temperature are an ideal medium for mesophilic growth. Control can be maintained over this group of bacteria by implementing cold or hot temperature treatments.

THERMOPHILES

Thermophiles are heat-loving bacteria, which grow at temperatures between 113°F and 150°F. Proper canning (240°F at 12–15 psi) kills these bacteria, but consuming

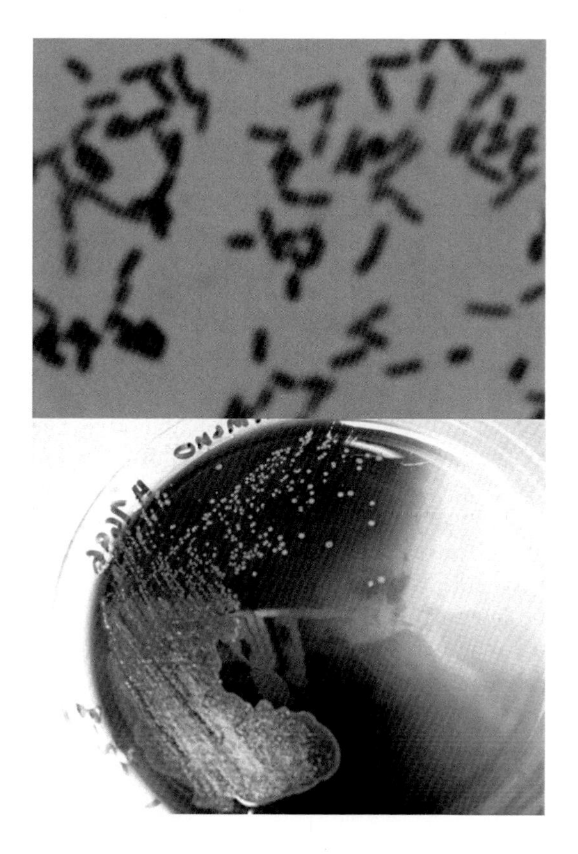

FIG. 12.7

Examples of *Listeria monocytogenes* cells and colonies on Modified Oxford Agar (MOX) plates.

Courtesy of Dr. James Dickson, Department of Animal Science, Iowa State University.

the products in which bacteria survive in improperly cooked or canned meats causes illness and in some instances can cause death. Some bacteria that cause illness in humans can grow in this temperature range. The current USDA FSIS cooling (stabilization) regulations for cooked meats are designed to minimize the growth of these pathogens.

SOURCES OF BACTERIAL CONTAMINATION

Bacteria are ubiquitous—they are everywhere. They can be found from the farm to the family table. The entire world, including food, is not sterile. It requires vigilance and meticulous efforts and actions in all areas of meat production, processing, sales, and consumption to control and kill the bacteria. Consequently, there are many sources of bacterial contamination in meat. They can be found in air, soil, and water. Soil is full of bacteria, and they can be spread by air currents. From the air, bacteria fall to contaminate soil and water. Bacteria can be found in crops and animal feeds that make them carriers of bacteria in their intestinal tracts. Feces then become a source of these microorganisms. Moreover, many kinds of bacteria are found in the human intestinal tract.

Animals, whether grazing in pastures or produced in feed lots and pens, are subject to many bacteria. As a result, the environment of the animal may play a major role in bacterial contamination of our eventual food supply. Improper dressing and evisceration procedures, regardless of the environment, spread bacteria to the carcass.

Animals have a natural defense system against bacteria in or on them, but once they are harvested for food, they lose their blood and immune systems, both of which are important in protecting the health of the animal. When the animal goes through the harvest process, it can be contaminated by the sticking, blood removal, and dressing process, for example, dirty sticking and skinning knives, hides and skins, scald tanks, handlers, internal organs, especially intestinal tracts, and storage. In a healthy animal, skeletal muscle tissues are usually sterile, but once the carcass is cut into wholesale and retail cuts, bacteria from the carcass are spread on the surface of these cuts by handlers and equipment. Also, grinding meat to manufacture ground meat products results in uniform distribution of microorganisms throughout the product. Because of this distribution, it is essential to cook ground meat to 160°F internally to prevent food-borne sickness and disease. Particle size reduction by chopping and mixing of a meat product to facilitate the distribution of ingredients can also spread contamination. A major source of contamination of products, equipment, and people with pathogenic bacteria is usually fecal material. Floor drains, ceiling fans, packaging equipment, and packaging materials can also be sources of bacteria in a meat processing room. If processing equipment is not properly cleaned and sanitized, biofilms can form from the residual bacteria. Once a biofilm forms, the bacteria are much more difficult to remove or kill during sanitation.

Workers (food handlers from packing or processing to the retail markets or restaurants) who practice cleanliness play an extremely important role in keeping themselves, facilities, equipment, and products safe and wholesome. Workers need to be knowledgeable about cleanliness and practice wearing hair and beard nets, clean clothes, and gloves. Also, workers need to be aware of bacterial contamination coming from their skin, nasal passages, respiratory tract, sneezing and coughing, feces, and urine. Proper hand washing must occur after going to the bathroom. Workers must realize the significance of proper hygiene, especially for hand washing because dirty hands are a major avenue of spreading spoilage microorganisms and food-borne pathogens not only in food but to other people. The application of this critical aspect of sanitation cannot be overemphasized. Minimizing, controlling, and stopping the growth of microorganisms is a must in supplying clean, safe, wholesome, tasty, and healthful meat and meat products.

FACTORS AFFECTING BACTERIAL GROWTH

Meat serves as an excellent source of nutrients for bacteria and also provides a hospitable environment for their growth and proliferation. Why is fresh meat such an excellent growth medium for bacteria? The answer is simple. It has their nutritional and water requirements for growth and reproduction. That is, meat has all the essential amino acids, vitamins, minerals, and moisture to provide a superb growth medium for growth and reproduction.

pH VALUES

The pH value of fresh meat at pH 5.6 or higher allows substantial bacterial growth, and the higher the pH value, the greater the growth. On the other hand, fermented products such as salamis, cervelats, and summer sausage have a pH of 4.6 or lower, and this greatly inhibits bacterial growth. Specifically, spores of *C. botulinum* are not able to germinate at pH levels below 5.0. At higher pH values meat needs to be processed (canned) at temperatures of 121°C (250°F) to destroy these spores.

WATER AND WATER ACTIVITY (a_W)

The high water content and water activity of meat make it an ideal growth medium for bacteria. It needs to be pointed out that water content in meat and water activity (a_w) of meat are not the same. A number expressing the ratio of the vapor pressure of food to the vapor pressure of water for the growth requirements of microorganisms is defined as a_w. Fresh meat has an a_w of 0.98, well above the minimum requirements for growth of bacteria, yeast, and molds. Insofar as water activity for microbial growth, bacteria require an a_w of 0.88, yeasts 0.8, and molds 0.7. Consequently, most fresh meats and meat products have a sufficient a_w for all three classes of microbes. An exception to products with a high a_w is, for example, dry cured hams and jerky (see Chapter 13). These processed products have a very low water activity (a_w of <0.85) and prevent microbial growth in most instances. Some dry cured hams can show evidence of mold growth over long curing periods used in their production, and this growth is usually harmless in the dry cured products. However, preventing growth is not the same as killing the bacteria, as pathogenic bacteria such as *Salmonella* can survive in low a_w conditions. In the early 2000s, there were several *Salmonella* outbreaks associated with jerky, demonstrating the bacteria's resistance to drying of the meat products.

TEMPERATURE

Temperature, in combination with time, is a major means of preserving the sensory quality characteristics of color, aroma, and flavor of meat and preventing food-borne illnesses. Cold temperatures of −10°C to 0°C (14–32°F) still allow cold-loving (psychrophiles) and cold-tolerant (psychrotrophs) spoilage bacteria (Fig. 12.8) to grow,

Classification	Examples
Thermophillic	*Thermus aquaticus*
Thermotolerant	Most spore-forming bacteria (spores)
Mesophillic	*Escherichia, Enterobacter, Salmonella*
Psychrotrophic	*Pseudomonas*, lactic acid bacteria, *Listeria*

Thermophilic—grow above 55°C

Thermotolerant—will survive but not grow at temperatures above 55°C

Mesophillic—grow at moderate temperatures, 10–45°C

Psyrchrotrophic—grow at temperatures below 10°C

FIG. 12.8

Effect of temperature on bacterial growth.

Courtesy of Dr. James Dickson, Department of Animal Science, Iowa State University.

but slowly. *L. monocytogenes* (Fig. 12.7), however, can grow very well at 0°C (32°F). For self-service case meats, variation in cold temperatures can decidedly affect the rate of growth of cold-loving bacteria. Colder storage temperatures are beneficial for both fresh and processed products in that they inhibit bacterial growth.

A practical example of an effect of temperature at the retail meat case on beef steak quality is as follows. Beef steaks displayed at 32–35°F will have less bacterial growth, brighter cherry red lean color, and an acceptable (salable) shelf life for about 3–4 days. In contrast, a retail case operating at 38–42°F will have more bacterial growth on the steak surfaces. The higher bacterial growth will discolor the steak more quickly resulting in a brownish/greenish color from a bright cherry red color. As a result, the salable shelf life of the discolored steak is about 1 or 2 days. Longer shelf life is essential because not all beef steaks are sold in 1 or 2 days, and steak color is a major determinant of sales. Discolored steaks (Fig. 12.9) will have to be offered at lower prices or be reworked in another product or discarded resulting

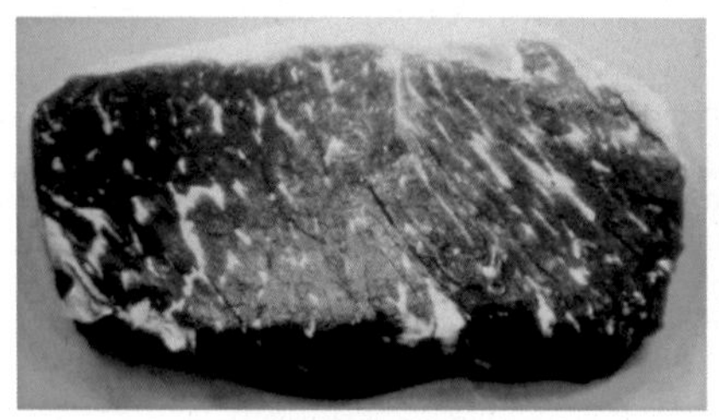

Discolored beef steak
Brownish-greenish color

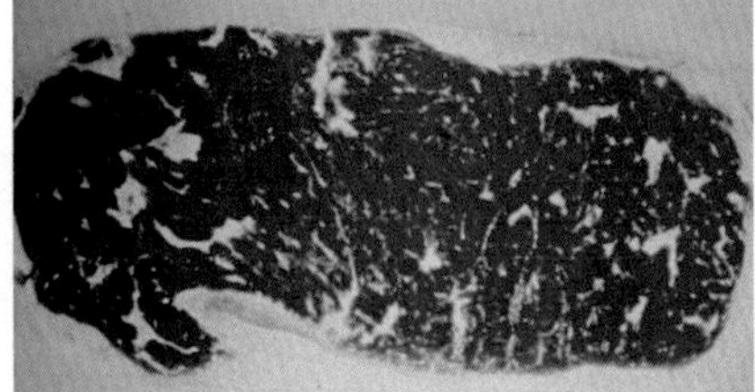

Normal-colored beef steak
Cherry red color

FIG. 12.9

Example of a beef steak discolored from bacterial contamination.

in loss of profitability. Salt-loving (halophilic) bacteria and lactic acid-producing bacteria found on cured meat products are also controlled by cold temperatures. It pays to keep the retail case cold! Normal bacterial populations on fresh meat cuts at refrigerated temperatures will be 10^3–10^5/cm^2 when placed in the meat display case, and steaks will remain eye appealing. With time or warmer temperatures, bacterial growth will accelerate and produce visible colonies and slime. Bacterial populations on meat that has reached the end of its shelf life, as defined by the consumer, vary considerably. Generally, populations in the high 10^6 to low 10^7/cm^2 indicate that the product is at the end of the shelf life. Meat with bacterial growth characterized as slime will have populations as high as 10^8/cm^2. Accompanying slime production is the putrid odors from the degradation of meat by bacteria.

Food-borne pathogens, classed as mesophiles (Fig. 12.8), grow best between 59°F and 122°F. Either temperatures below or above this range will control growth of the pathogens. An example of thermophilic (heat-loving) bacteria is those found in improperly canned meats. If the canned product has not been heated to 170°F, or above, the thermophilic bacteria will cause spoilage and illness and depending on the level of contamination, even death. *C. botulinum* (see Fig. 12.5), an anaerobe, is the spore-forming pathogen commonly found in improperly heated and canned foods.

Below 14°F bacteria cells become dormant because of the bacteriostatic and bactericidal effects of decreased water availability due to freezing. Cold temperatures, above and below freezing temperatures, are essential for preserving meat, maintaining organoleptic properties, increasing profitability, and preventing food-borne illnesses.

Bacteria also have different oxygen requirements to proliferate and sustain their growth. The bacteria that require 20% oxygen or more for their growth are called aerobes. Common bacteria, such as *Pseudomonas*, need oxygen to grow on meat. There is a group of microorganisms that can grow in the presence or absence of oxygen termed facultative anaerobes (see Fig. 12.6). However, these bacteria generally grow slower in the absence of oxygen. Another type requiring oxygen is the microaerophilics. They need only 6% oxygen to meet their growth requirements because the 20% typically found in the atmosphere is toxic to them. The fourth type is anaerobes (Fig. 12.6), and they can grow without oxygen. For desirable fresh meat color, oxygen must be present, but for longer storage before retail display, some cuts are vacuum packaged, which can reduce bacterial growth and extend the shelf life and the eating quality characteristics of meat by excluding oxygen. Hence, the more common spoilage bacteria and pathogenic bacteria are prevented from growing when oxygen is excluded from the package.

BENEFICIAL MICROORGANISMS

Not all microorganisms cause spoilage, illness (disease), or death. Some bacteria, molds, and yeast are beneficial. For example, bacteria such as *Lactobacillus* are important in fermentation of sugar to lactic acid of processed meat products such as salami, pepperoni, and summer sausage to produce the low pH and tangy taste. The development of the characteristic flavor of the product and protection against spoilage obviously are very valuable. Yeasts are useful in the fermentation of sugar to alcohol. And molds are associated with country cured ham and some mold-ripened cheeses, such as blue cheese.

Another area of benefit to the meat industry and consumers is that certain bacteria are competitive inhibitors of other bacteria on fresh meat surfaces. These are found naturally on fresh meat cuts to decrease the growth of other less desirable spoilage bacteria causing shelf life to be longer. As a result, correct use of competitive inhibitors allows meat to have a brighter color and reduce the chances of food-borne illnesses.

Natural antimicrobial-produced materials from safe bacteria can be added or may already be in a product to inhibit the growth of spoilage or food-borne pathogens. They are termed bacteriocins and occur in nature. These are produced by safe bacteria and inhibit the growth of other harmful bacteria, allowing the good bacteria to survive. Use of substances produced by microorganisms, or those occurring naturally in meats to minimize spoilage and increase safety from food-borne pathogens, is being experimentally tested. These can be added to meat products without adversely affecting sensory, nutritional, and safety features of the product when added in the appropriate quantity.

PATHOGENIC BACTERIA

Pathogenic bacteria are those microorganisms capable of producing diseases causing illness and even death. These bacteria are largely responsible for outbreaks of food infections and intoxications, and thousands of deaths can occur from these bacteria. Most of these bacteria are classified as mesophiles. That is, they grow well between 68°F and 122°F. As a result of the consumption of foods contaminated with pathogenic microorganisms, the cost of food-borne illnesses is astronomical. Economically, millions of dollars are spent annually on medical costs, lost time from work, and insurance. These costs from consumption of contaminated meat and meat products result from over a third of the total outbreaks of food-borne illnesses. Consequently, it is imperative to understand the importance of food-borne microbes and their control in the production of clean, healthful meat products. The following is a brief description of the major bacteria causing infections, intoxications, and death.

CLOSTRIDIUM BOTULINUM

This is one of the most toxic food-borne pathogens. It is found in soil, water, and food and lives strictly in anaerobic conditions such as improperly canned products. It is capable of producing endospores (see Fig. 12.5), and it is the ingested toxin that causes botulism, a food-borne intoxication. It often occurs between 12 and 72h after consumption of contaminated meat. It is a powerful neurotoxin causing headache, double vision, difficult breathing, and death due to paralysis of involuntary muscles and respiratory failure. The only cure is an antitoxin and supportive care. This pathogen grows easily in the absence of oxygen and is associated with underprocessed canned products. Nitrite in processed meat products inhibits germination of *C. botulinum* spores. Botulism is an extremely rare but potentially dangerous disease in the United States.

CLOSTRIDIUM PERFRINGENS

The *C. perfringens* bacteria are spore formers and are found in sewage, dust, air, and the intestinal tracts. If people consume contaminated meat, the bacteria produce a toxin in the small intestine. These spore formers are anaerobic-aerotolerant. The lowest temperature for growth is 68°F, and sporulation requires 98–99°F. Cooked meat should be kept above 135°F or below 50°F to stop growth. Rapid cooling is essential, and the current USDA stabilization regulations are based on preventing the growth of this bacterium. These food-borne bacteria are associated with improperly cooked and stored meats. The result is nausea, vomiting, diarrhea, gas pain, and abdominal cramps in 8–24 h. Death from poisoning by this food-borne microbe is rare.

STAPHYLOCOCCUS AUREUS

S. aureus (Fig. 12.10) is a gram-positive bacterium and can be found on skin, hands, and in nasal passages and wounds. It is spread by sneezing, coughing, skin irritations,

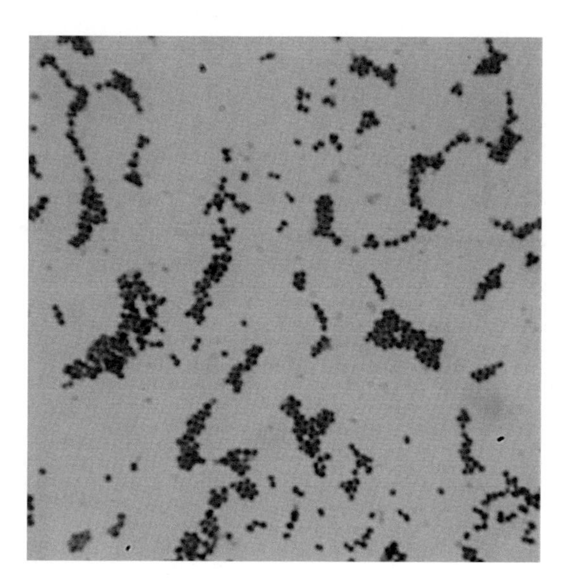

FIG. 12.10

An example of *Staphylococcus aureus* bacteria.

Courtesy of Dr. James Dickson, Department of Animal Science, and Robert Hubert, Interdepartmental Program, Iowa State University.

and direct contact of the skin with foods. It is people associated, and food is simply the growth medium. *S. aureus* is a facultative anaerobe that produces a heat-resistant toxin and can grow at low a_w (0.87). It can be found in cured meats; it tolerates both salt and nitrite. The symptoms of ingestion of the toxin of *S. aureus* are nausea, vomiting, and diarrhea within a few hours after ingestion of the toxin. Proper heating and cooling of foods will prevent staphylococcal intoxication.

SALMONELLA SPP

A commonly found pathogen is *Salmonella* (Fig. 12.11). By consuming *Salmonella*-contaminated foods, one can get a food infection called salmonellosis. This is caused by ingestion of the live bacteria from improperly cooked meat and mishandled foods. Symptoms include headache, diarrhea, fever, vomiting, chills, and abdominal discomfort within 12–72h and may last as long as 4–7days. It is usually resolved in 5–7days. It is a very common infection in the United States affecting over one mil-

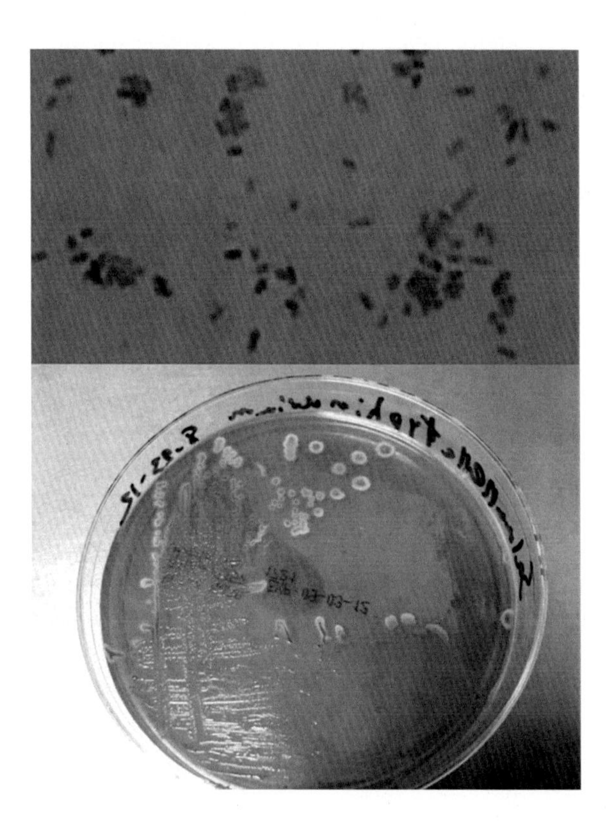

FIG. 12.11

Examples of *Salmonella typhimurium* cells and colonies on Xylose Lysine Decarboxylase Agar (XLD) plates.

Courtesy of Dr. James Dickson, Department of Animal Science, Iowa State University.

lion people each year with about 550 deaths of acute salmonellosis. Raw foods of animal origin such as beef, poultry, milk, and eggs are common vehicles. It is found in intestinal tracts of humans and animals, especially poultry. Poultry meat has a higher incidence of *Salmonella*, although all meat products have seen a dramatic decrease in the incidence of *Salmonella* since the late 1990s. Adequate cooking (it is easily destroyed by cooking) and cleanliness will usually prevent its growth.

CAMPYLOBACTER JEJUNI

C. jejuni (Fig. 12.12) is a gram-negative bacterium causing more food and water infections than any other bacterium. This dangerous food-borne pathogen is found in soil, water, humans, animals, and food, and it is transmitted by ingestion of raw meat, milk, or contaminated water. In the ideal environment of the intestine, it multiplies and produces a cytotoxin in 2–8 days that results in vomiting, severe diarrhea, including blood diarrhea, and ultimately dehydration. It has interesting growth characteristics as it is microaerophilic—it needs 6% oxygen, but 20% oxygen is toxic to it. It grows best at 107°F. Adequate cooking and cooling of foods prevent illness. This bacterium has developed resistance to most antibiotics and therefore can be a major concern for the food industry in the future.

LISTERIA MONOCYTOGENES

This pathogen was almost unheard of until the 1980s. *L. monocytogenes* (Fig. 12.7) is found everywhere and has a wide range in growth temperature (35–113°F) and pH (5.0–9.6). Because of these wide ranges it becomes difficult to control. Because of *L. monocytogenes* contamination, there have been many recalls and outbreaks of

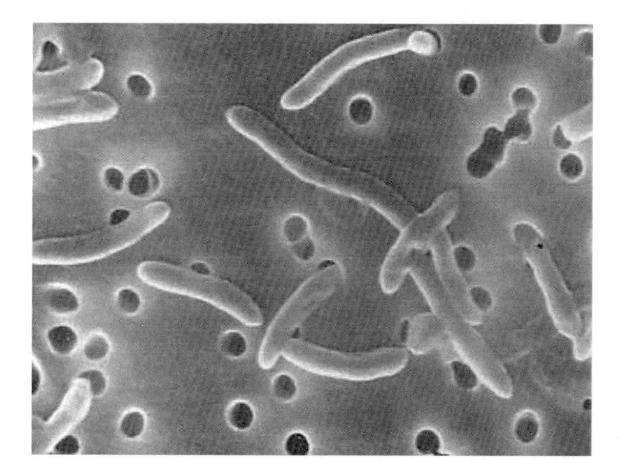

FIG. 12.12

An example of *Campylobacter jejuni* bacteria.

Courtesy of Dr. James Dickson, Department of Animal Science, Iowa State University.

ready-to-eat processed meat products. When *L. monocytogenes* is consumed, susceptible humans get an infection called listeriosis. Symptoms are similar to flu—fever, muscle aches, nausea, headache, upset stomach, and vomiting. Time of onset can be 3 days to weeks. About 2000 cases of listeriosis and >400 deaths are reported annually in the United States. Serious listeriosis infections result in higher deaths than either *Salmonella* or *Escherichia coli*, and two of the four deadliest food-borne disease outbreaks in the United States were attributable to *Listeria*. The susceptible population is primarily the young, pregnant, cancer victims, elderly, and immune-compromised patients.

L. monocytogenes, a gram-positive rod, is found in the intestinal tract and fecal content of humans and in the environment. Good sanitation practices are imperative throughout the meat processing facilities because many elements such as soil, sewage, dust, water, fish, sausage, and vegetables contain *L. monocytogenes. L. monocytogenes* is often seen as a postcooking contaminant, where the product becomes contaminated from an environmental source after the smokehouse processing and before packaging. It is unique in that it can grow on products at refrigerated temperatures and even survive freezing. It is also salt tolerant and resists nitrite and acidity. Organic salts such as sodium diacetate, potassium benzoate, and sodium lactate effectively inhibit the surface growth of *L. monocytogenes* on frankfurters. Pediocin, a bacteriocin, has been shown to be effective in prolonging shelf life and reducing growth on frankfurters. Use of proper cookery, strict sanitation, and clean handling prevent its growth and the food illness. Meat must be cooked to 160°F internal temperature to prevent infection.

ESCHERICHIA COLI

This is the most common bacterium found in human and animal intestines. There are hundreds of strains, most of them harmless and living in the intestines; however, there are a few very harmful strains, of which *E. coli* O157:H7 (Fig. 12.13) is perhaps the best known. It is a strain of *E. coli* and a member of the enterohemorrhagic *E. coli* group. It is associated with hemorrhagic colitis resulting in bloody diarrhea, nausea, and severe abdominal pain and can cause death. Symptoms occur easily and one can become infected from undercooked contaminated meat, raw produce, or an infected person not washing their hands after a bowel movement or changing a diaper. *E. coli* easily spreads from animals to people or raw or undercooked meat products via fecal contamination. Fecal contamination can occur in ground beef because of the fecal contamination on hides during the dressing process that can be transferred to the carcass and then into the product during grinding. This illness is possible by consuming improperly handled and cooked meat products, especially ground beef. Insufficient cooking of ground beef can set off national publicity, sickness, and death by the consumption of these undercooked burgers. Burgers cooked to 160°F internal temperature can prevent the effects of *E. coli* O157:H7, and it is worth noting that there has not been an outbreak of *E. coli* O157 traced to a major national quick-serve restaurant since 1995. Proper dressing procedures, sanitation, and cookery prevent *E. coli* O157:H7 from contaminating food and causing illness or death.

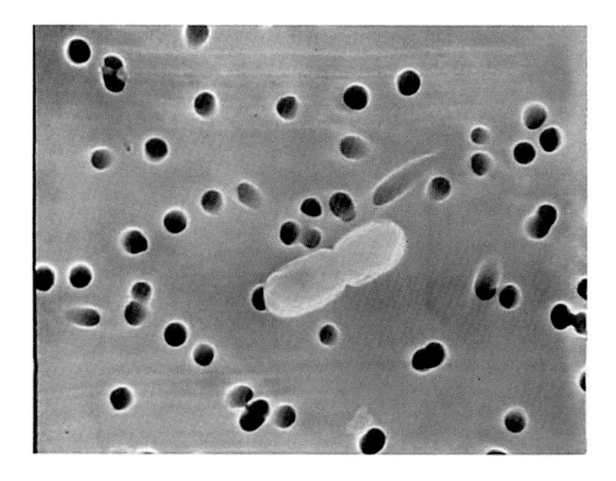

FIG. 12.13

An example of *Escherichia coli* O157:H7 bacteria.

Courtesy of Dr. James Dickson, Department of Animal Science, Iowa State University.

Other bacteria such as *Shigella* spp., *Bacillus cereus*, and *Yersinia* spp. are also food-borne pathogens of concern but are not as prevalent or powerful as the other aforementioned pathogenic bacteria.

PREVENTING SPOILAGE AND ILLNESS

The basic needs of bacterial contamination of meat are the source of contamination, the proper environment, the appropriate temperature to grow, and the sufficient time for growth. It all begins with cleanliness on the farm, in the plant, and in the home to reduce the sources of bacteria and prevent meat spoilage and contamination by food-borne pathogens. Cleanliness is the basis for decreasing, eliminating, and controlling bacteria on animals, meat products, equipment, packages, facilities, and people. For example, for cleaning systems to be effective they must have 140–180°F water and steam. In addition, appropriate detergents and sanitizers are essential for cleaning facilities, equipment, and clothes. Furthermore, proper sanitation practices must be applied, and therefore an individual or team of people who clean meat facilities must be thoroughly knowledgeable about the importance and effect of cleanliness.

The positive benefits of sanitation on human health and well-being are immense. The cost of food-borne illness because of the consumption of contaminated meat can be in the billions of dollars. Time lost on the job and medical and insurance costs are astronomical. In addition to the well-being of people, cleanliness will increase the shelf life of meat products, retain the organoleptic quality, reduce product loss, and enhance public image. All of these taken together mean more profitability. Sanitation pays and is an essential component of any food processing operation! Indeed, proper hand washing is a necessity!

To ensure product quality and safety, proper temperatures must be applied. Steam pasteurization of carcasses done shortly after the meat animal harvest process results in reducing bacterial contamination of beef carcasses. After harvest, rapid chilling of carcasses and cold storage temperatures are required. Storage temperatures below 41°F are essential in slowing spoilage and stopping pathogenic bacterial growth. Even at refrigeration temperatures, cold-loving bacteria will grow at 28–40°F. Below freezing, 28°F and lower, bacterial growth will be slowed (bacteriostatic effects), and below 0°F with sufficient time bacteria will actually be killed (bactericidal effects) on meat. Proper defrosting of frozen meat is essential in preventing food illnesses. Frozen meat should be thawed in a refrigerator or microwave oven to avoid growth of microbes that would otherwise grow if meat was thawed at room temperature.

Insofar as temperature is concerned for foods to be consumed, keep hot foods hot and cold foods cold. As with bacterial growth, the temperature and time interact to destroy the microbes; higher cook temperatures require shorter cook times, while lower temperatures require longer times. Proper cooking to 155°F internally is essential in destroying pathogenic bacteria. Salmonellae are destroyed at about 160°F as well as almost all nonspore-forming bacteria. To destroy potential *E. coli* O157:H7 in ground beef, it must be cooked to an internal temperature of 160°F. The reason steaks and roasts can be cooked to a lower internal temperature is because of the sterile interior of muscle. Time as well as cooking temperature is important in the destruction of bacteria. A longer cook time is required for roast beef cooked at 135°F internal temperature to kill *Salmonella* compared with instantaneous destruction of it at 145°F internal temperature. Leftovers should be promptly refrigerated or frozen after staying out no longer than 2 h at room temperature. By leaving meat at room temperature, rapid growth of bacteria occurs and allows the potential development of off-flavors and food illnesses. Very high temperature above 240°F at pressures of 12–15 psi will sterilize (no microbial survival) a meat product. Consequently, proper retort canning of meat in hermetically sealed containers will preserve a product indefinitely.

Bacteria, indeed all microbes, can be controlled by lowering water activity. A case in point is a dried processed product like beef jerky. The a_w is so low (0.85) bacteria cannot grow even at room temperature. Also, fermentation of a product by producing lactic acid to drop the pH to 4.6, like in a summer sausage, will not allow bacteria to continue to grow. Adding salt will slow bacterial growth or even inhibit it in high-salt meat products such as country cured hams, and adding nitrite in cured meat products inhibits the growth of *C. botulinum*.

Removing oxygen from packages of meat will control the growth of aerobic bacteria. Elimination of oxygen is why vacuum packaging plays a key role in extending the shelf life of meat products. Smoking of meat products is another way of preventing spoilage. The chemicals in smoke, such as phenols, also inhibit bacterial growth.

Irradiation of foods by using ionizing radiation, either electronically or isotopically produced, will greatly reduce but not eliminate all bacteria. Low-dose irradiation of 1 kilogray (KG) is effective in eliminating many spoilage microorganisms and as a result extends shelf life under refrigerated conditions. Irradiation is especially effective in ensuring the safety of ground beef. Some commercial applications of irradiating ground beef are used effectively.

The use of the "hurdle" concept of preventing bacterial growth is valuable. It uses a combination of steps to inhibit or destroy bacteria. For example, the use of nitrite to inhibit sporulation, salt to lower the a_w, followed by heat treatment, and then storage at refrigerated temperatures in combination are much more effective than any one of these methods alone in the preservation of meat products.

METHODS OF ENUMERATION

Conventional methods of plate counts are still being used for enumeration, but with advanced technology, new methods are becoming rapid, convenient, and simple. The conventional method is the total aerobic plate count. Meat samples are homogenized in a buffered solution, placed onto a Plate Count Agar, and the colonies (Fig. 12.14) are counted after they have been incubated at 95°F for 24–48 h. This method tells how many but not what kind of bacteria.

To give more precise information about the bacteria present on the food product, selective agars are used to eliminate bacteria that are not of interest for culturing. Chemicals are added to inhibit the growth of other bacteria and allow the ones in question to grow. Several biochemical tests are also commercially available to identify particular classes of microbes. Also, assays are available to detect the presence of bacterial cells or toxins in the meat by using immunochemical techniques. An antibody specific for the organism is used in an agglutination reaction to detect its presence.

The Enzyme Linked Immunosorbent Assay and the DNA Probe Assay are the most advanced techniques and are commercially available. Detection of pathogens such as *Salmonella* and *Listeria* are possible and can be accomplished in a matter of a couple of hours, after a suitable preenrichment step. The shortest time frames from beginning the sample analysis to an end result are still typically in the 48- to 56-h time range. These methods are more costly than conventional methods in terms of materials, but the labor savings and rapid results offset the cost.

AGENCIES, ORGANIZATIONS, AND INFORMATION FOCUSED ON THE SAFETY OF MEAT AND MEAT PRODUCTS

USDA

The United States Department of Agriculture (USDA) has been the guardian of much of our food supply for many years. It dates back to 1906 when the Federal Meat Inspection Act was passed. Between then and now a number of modifications and improvements have been passed and implemented. A significant agency of the federal government for the purpose of protecting our meat supply is the

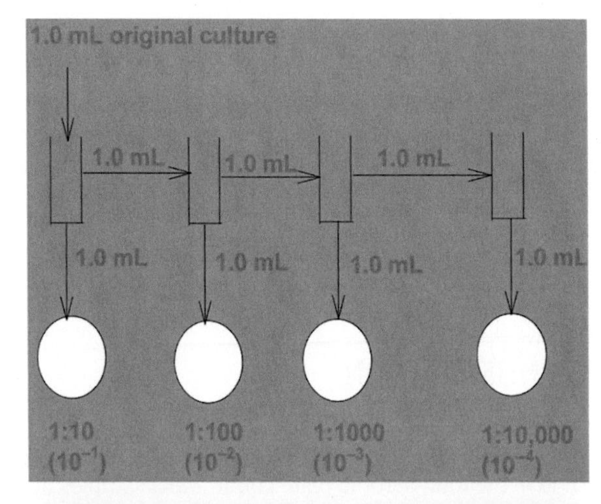

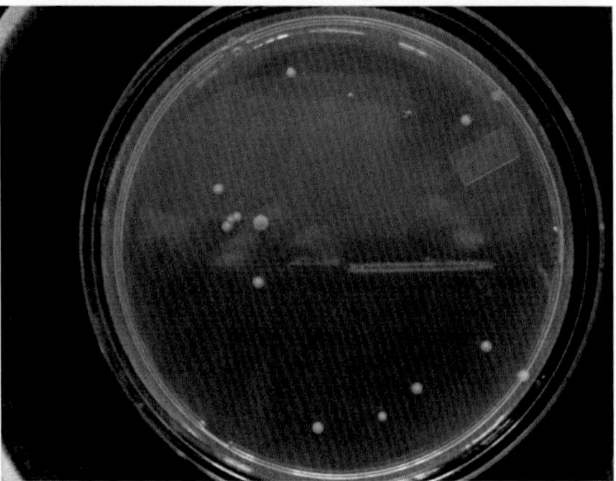

FIG. 12.14

An example of agar plate counts.

Courtesy of Dr. James Dickson, Department of Animal Science, Iowa State University.

Animal and Plant Health Inspection Service (APHIS). It works to protect the public from livestock diseases transmissible to man and conducts nation-wide surveys for specific microorganisms in production livestock systems. It also prevents and eradicates livestock diseases that are economically destructive to farmers, and it prevents foreign diseases from entering the United States. Another significant agency, Food Safety and Inspection Service (FSIS), is responsible for seeing that meat and poultry products moving in interstate and foreign commerce are safe, wholesome, unadulterated, and are accurately marked, labeled, and packaged for use as human food. Obviously, FSIS has immense responsibilities in guaranteeing meat wholesomeness, safety, and accurate labeling. US citizens through their taxes pay for meat inspection.

HACCP

The implementation of HACCP, beginning in 1996, has improved food safety by applying scientific principles to prevent meat contamination, especially giving strong focus on pathogenic bacteria awareness. HACCP specifies the hazards, shows their likely location, calls attention to the critical control points, and takes the appropriate action to manage the process. Companies are vigorously carrying out these principles to help ensure safe, wholesome meat and meat products. By any measure, the national implementation of HACCP has been a success for the industry and most importantly for the consumer.

Along with the implementation of HACCP, FSIS has developed *Salmonella* performance standards (legally mandated) for beef, swine, and poultry in an effort to more quickly report testing results to target establishments needing improvement and provide timely information to both industry and consumers. Along with *Salmonella* control, there is a continued focus on beef for *E. coli* O157:H7 contamination control by the industry.

Quality control (assurance) personnel must work hand in hand with HACCP objectives to ensure safe, wholesome meat. They also must focus on efficiency, service, and profitability of a meat operation by following various protocols. In recent years, meat plants and processors have hired food safety managers and staffs to effectively control all points along a meat production and processing line. As a result, a serious situation from contamination with pathogenic microorganisms, chemical residues, or physical objects is avoided.

From a state government agency perspective, the Bureau of Meat and Poultry Inspection is involved in regulating animal harvest and meat processing operations. Other organizations such as livestock and meat commodity groups, university extension personnel, and local health departments are valuable sources of personal information about safe food handling and food-borne illnesses.

For consumers at the retail level, safe-handling labels on raw meat and poultry products provide basic guidelines in keeping meat safe. These guidelines are as follows: (1) keep refrigerated or frozen; (2) thaw in refrigerator or microwave; (3) keep raw meat and poultry separate from other foods; (4) wash working surfaces (including cutting boards), utensils, and handles after touching raw meat and poultry; (5) cook thoroughly; (6) keep hot food hot; and (7) refrigerate leftovers immediately or discard.

SUMMARY

The prevention of spoilage and food-borne illnesses in meat and meat products is critical to healthful living and profitability. Bacteria are the main cause of meat spoilage and food-borne illnesses, and they are everywhere. Consequently, safety measures must be implemented from the farm to the family table. Proper use of sanitation, temperature, and time to ensure safe consumable refrigerated and cooked products must be practiced. Strict adherence to plant and quality control standards and government regulations is fundamental in effectively preventing meat spoilage and food-borne illnesses. Cleanliness of industrial personnel, equipment, facilities, meat handlers, and the home environment is a major step in providing safe, wholesome, attractive, and delicious-tasting meat and meat products. A teamwork approach and a strong effort by government, academia, industry, and consumers are essential in maximizing meat safety.

QUESTIONS FOR STUDY AND DISCUSSION TOPICS

1. What percentage of disease outbreaks in the United States are associated with contaminated meat?
2. Name the three types of hygiene control that must be diligently implemented to help ensure a safe, wholesome meat supply.
3. Define a microbe, and list where they may be found.
4. Define a bacterium, and give the size of a typical bacterium.
5. Describe the function of spore-forming bacteria when they are in an environment not conducive to normal growth.
6. *L. monocytogenes* is a significant pathogen. What type of meat products are often contaminated with this bacterium? Can they grow at refrigerated temperatures?
7. What are the major sources of bacterial contamination in meat and meat products?
8. List the four major factors affecting bacterial growth.
9. Are some microorganisms helpful for developing special meat products? Provide an example.
10. List three microorganisms that have the most influence on producing deaths in people from eating meat or meat products in the United States.

CHAPTER

Fresh and cured meat processing and preservation

13

INTRODUCTION

Processed meat and meat products are manufactured from meat obtained from the carcass and edible offal (noncarcass parts such as liver, kidney, and heart). Most of the processed fresh meat is prepared by high-speed and high-technology equipment, but some special products are still prepared by the traditional methods used for many years across the globe. An example would be the dry-cured ham. Fresh processed meats are a very large part of the meat marketing system in the United States. Ground beef is classified as a processed fresh meat product so all of the fast-food industries that use ground beef contribute to the future of this industry. The ingenuity and creativity of personnel in the processed meat industry result in new products on a regular basis to meet the consumer demand for fresh processed meat.

EXAMPLES OF PROCESSED FRESH MEAT

GROUND BEEF

Ground beef is a major market for beef in the United States. More than one half of the beef sold in the United States is in the form of ground beef. It has a big market share in fast-food restaurants, traditional restaurants, institutions, and in US homes. The terms ground beef and chopped beef are considered to have the same meaning. Ground beef is prepared by the use of mechanical, high-speed grinding and/or chopping of boneless beef cuts and trimmings. The manufacture of ground beef products is regulated by the United States Department of Agriculture—Food Safety and Inspection Service (USDA-FSIS) codes in which composition and labeling regulations of ground beef products are spelled out in detail. These regulations specify that ground beef must be made from fresh and/or frozen beef, with or without seasoning, and without the addition of fat, and is limited to 30% fat. Many ground beef products are much leaner (e.g., 90% lean, 10% fat, and they must be labeled as such). Furthermore, the regulations state that ground beef may not contain added water, extenders, or binders and not exceed 25% cheek meat (the masseter muscles of the head). Ground beef made from the round or ground beef made from the chuck must be listed on the package label to denote the cut or part used for making that specific product. Hamburger is a popular term used for ground beef, and the USDA definition

The Science of Animal Growth and Meat Technology. https://doi.org/10.1016/B978-0-12-815277-5.00013-5

for hamburger is only slightly different from that for ground beef. Based on its legal definition, hamburger can have added beef fat. Interestingly, hamburger has nothing to do with the pork carcass wholesale cut, ham.

A product labeled "beef patties" is different from ground beef in that beef patties can contain binders and extenders and may or may not have added water. The word patty is commonly used to describe ground beef products. Low-fat beef patties are those products combining meat and other nonmeat ingredients for the production of low-fat meat products. These products must be labeled as low fat, fat reduced, and/or containing nonmeat ingredients. In addition to being used in patties, it is also used in the manufacture of foods such as pizza, spaghetti, tacos, and burritos, and often it is frozen for use in ready to heat and serve dishes such as casseroles. A large amount of patty manufacturing takes place by using a continuous system of grinding, blending, forming, freezing, and packaging. Large beef-patty processing plants have equipment capable of producing 10,000 pounds per hour. Fat content is monitored online by rapid analytical methods like infrared (Fig. 13.1). Immediate and constant analysis is essential in producing the desired blends of lean and fat to meet company specifications and making necessary blend adjustments. Special meat grinder plates are available to greatly reduce or eliminate any bone particles that have been part of the beef trimmings (Fig. 13.2). Use of rapid cryogenic freezing substances such as liquid nitrogen (−80°F) has become more common because of its beneficial effects on decreasing cooking loss and improving the flavor of the ground beef products.

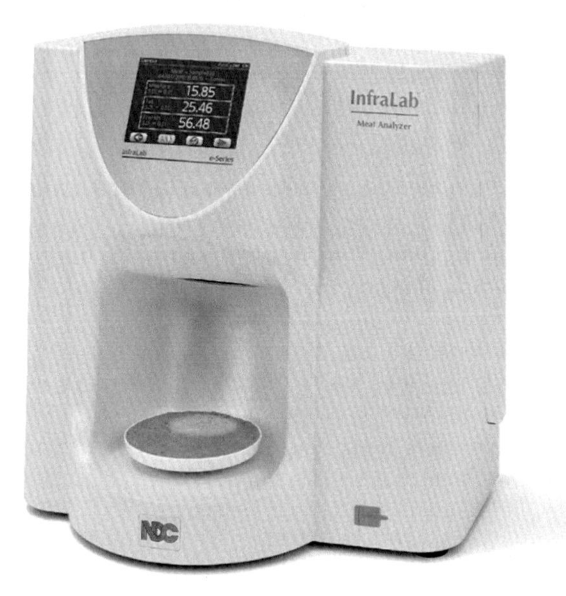

FIG. 13.1

An example of a rapid analytical method to determine the fat, moisture, and protein percentages of ground beef using an infrared unit.

From: NDC Infrared Engineering Inc., Irwindale, California.

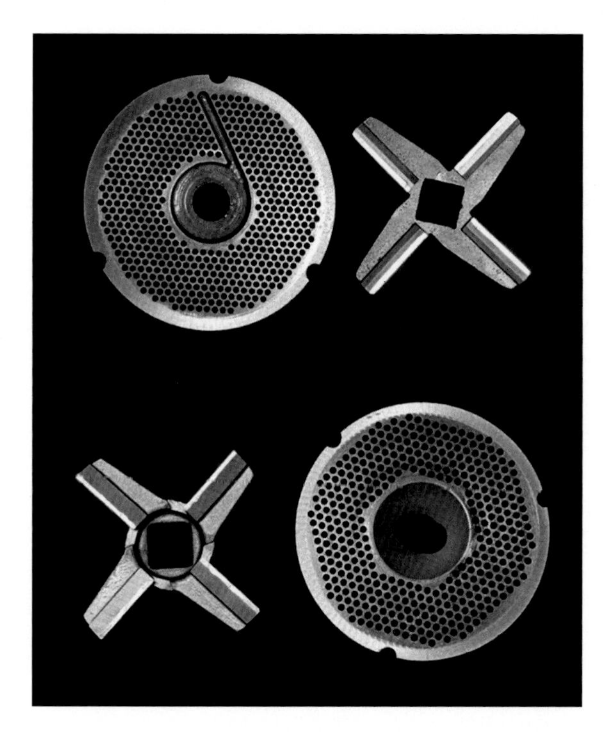

FIG. 13.2

Special meat grinder plates used to greatly reduce or eliminate bone particles in ground beef products.

Courtesy, Iowa State University Meat Science Laboratory.

After rapid freezing, the packaged ground beef product is placed in freezers for storage for subsequent shipment to retail stores, restaurants, and institutions.

Precooking patties at the wholesale level is becoming increasingly more popular because of the demand for rapid meal preparation and service, especially in the fast-food industry. Usually, the three stages of precooking doneness are fully cooked, partially cooked, and char-marked. Also, ground lamb, pork, chicken, and turkey patties are manufactured for retail sale using the same techniques described for beef patties.

MECHANICALLY SEPARATED MEAT

When meat is separated from the bone, there is still some muscle left on the bone. This meat can be recovered and used for meat processing. The mechanical process to separate muscle from bone starts by grinding the meaty bones from carcasses representing red meat, poultry, and fish. The grinder contains a sieve designed to remove bone particles and allow the meaty portion to pass through the sieve. Mechanically separated meat (MSM), mechanically deboned poultry, and minced fish are safe, nutritious, economical, and palatable. Another example would be finely textured lean. These products have wide usage in processed meat and contribute significantly to the world's protein supply.

Government regulations by USDA and the Department of Commerce spell out the limits on bone particles, calcium, and protein content; biological value; and the amount of the MSM allowed in a processed meat product. Mechanically separated red meats can be used in beef patties, pressed and chopped ham, sausages, stews, and spreads. It cannot be used in baby food, hamburger, ground beef, fabricated steaks, barbequed meat, corned beef, cured pork products, beef with gravy, or meat pies. When MSM is used in the marketplace, it must be labeled as Mechanically Separated and associated with the name of the species (e.g., Mechanically Separated Beef or Mechanically Separated Pork). Some MSM also goes into pet food. Mechanically separated poultry products can be used in baby foods, poultry rolls, structured products, and in all poultry sausages such as frankfurters and bologna. Mechanically separated poultry is limited to 15% in red meat franks and bologna.

MECHANICAL AND ENZYME TENDERIZATION

Another form of converting fresh meat into a more palatable product is the process of mechanical and enzyme tenderization. Lower USDA grades of fresh beef cuts (lower than USDA Select) are a major source of these tenderized steaks. Blade tenderizing machines are used to sever connective tissue and muscle fibers of boneless steaks or subprimals to produce more tender steaks (Fig. 13.3). Enzyme tenderization of steaks is accomplished most effectively by injecting steaks with a solution usually containing a proteolytic enzyme such as papain, which is obtained from the papaya plant, and bromelain from pineapple, at the meat processing plant. Improved tenderness comes from structural disruption by enzyme degradation of collagen and myofibrillar proteins when the steaks are cooked. Unfortunately, consumers often find wide variation in tenderness of mechanical- and enzyme-tenderized steaks.

FIG. 13.3

An example of a blade tenderizing machine.

Courtesy, Iowa State University Meat Science Laboratory.

This results in steaks that are not always acceptable in tenderness and flavor. In some cases, plant proteases, such as actinidin from the kiwi fruit, are also effective at improving tenderness, but actinidin is an allergen. It is for this reason that inclusion of nonmeat ingredients must be declared on the meat label.

RESTRUCTURED MEAT PRODUCTS

Restructured meat products are ground, flaked (Fig. 13.4), or chopped and manufactured into steaks, chops, or roast-like products for retail consumers and institutional food preparation. These products have a texture more closely identified with an intact steak or chop than that of a ground product. After particle size reduction, the meat is mixed with salt, phosphate, and protein materials (e.g., egg albumen) or hydrocolloid binders such as alginate, a gum extract from brown seaweed. Salt and phosphate act to solubilize myofibrillar proteins to ensure a stable meat product bind, whereas protein materials and carbohydrate binders serve in retaining the product structure. Then the product is formed into its desired shape, cooked, and packaged for retail

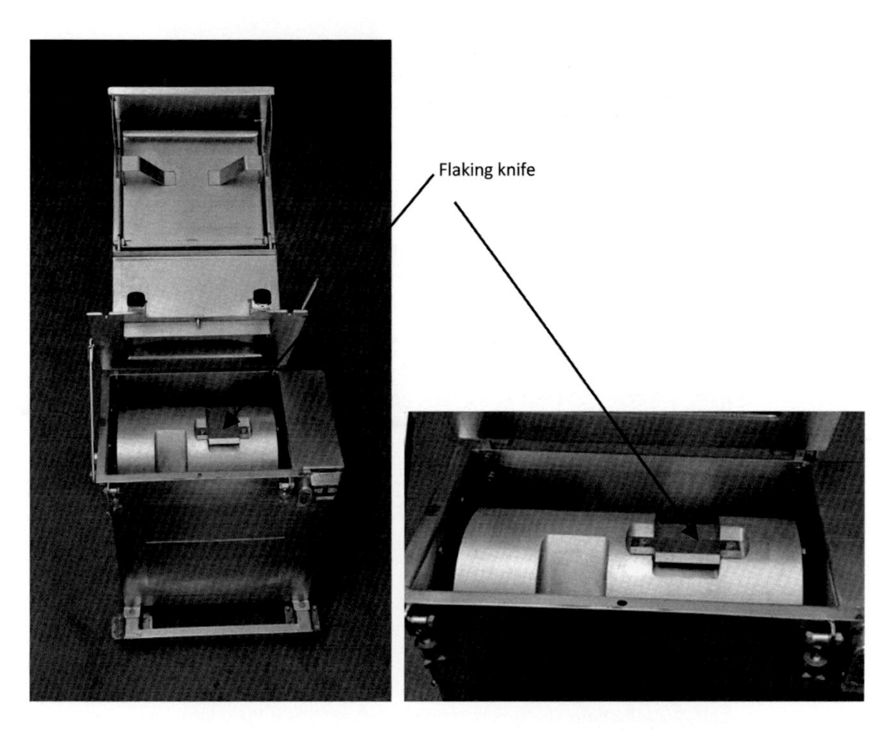

FIG. 13.4

A machine used to prepare flaked meat before it is manufactured into steaks or chops.

Courtesy, Iowa State University Meat Science Laboratory.

markets. The beauty of restructuring is that it affords the use of quality meat that can be transformed into even more valuable products (Fig. 13.5) by the processor. A large percentage of turkey and chicken meat is manufactured into restructured products to form loaves or roasts (Fig. 13.6) and other convenience products. For instance, breast meat is transformed into turkey rolls, steaks, and nuggets (Fig. 13.7).

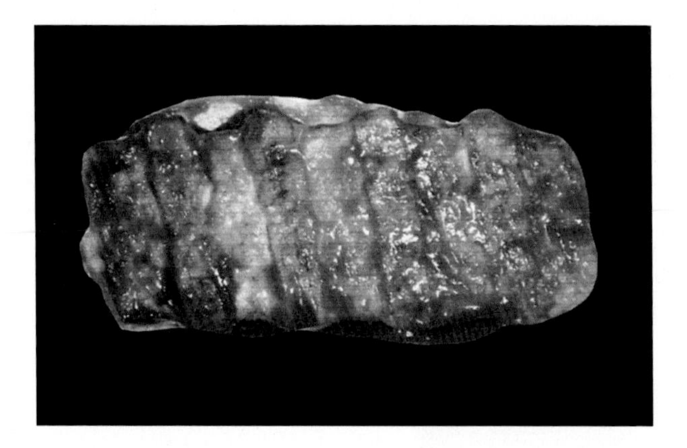

FIG. 13.5

An example of a restructured pork rib made from boneless pork cuts such as the pork carcass shoulder.

FIG. 13.6

An example of a roast manufactured from chicken breast meat and bacon.

Courtesy of Smithfield Foods.

FIG. 13.7

Turkey slices or steaks obtained from a quality turkey roast made from portions of turkey breast meat that was transformed into a roast.

MOISTURE- AND MARINATION-ENHANCED FRESH PORK PRODUCTS

The meat industry uses modern technology for injecting, massaging, and tumbling muscle to enhance tenderness, juiciness, and flavor of fresh meat products. These procedures are applied to improve product consistency, extend shelf life, and increase profitability. For example, pork loins are moisture enhanced by pumping the loins with a brine solution of 8%–12% and not usually >20% of "green weight." The term "green weight" refers to the weight of the meat by itself. The brine formulation consists of water, salt, and phosphates. It may also include flavorings (e.g., liquid smoke flavors) and preservatives, such as lactates and acetates. The products are vacuum packaged and refrigerated, and some products are cooked to 160°F internal temperature before they are shipped to retail markets. Marinades are also pumped into meat products from 20% to 30% of green weight. These marinade formulations include acid ingredients such as citric or acetic acid (vinegar), and flavorings and colorings such as teriyaki, honey mustard, garlic, barbeque, or peppercorn (Fig. 13.8). They are then vacuum packaged, chilled, and sent out to the marketplace. To produce quality moisture-enhanced and marinated products, the highest level of sanitation must be implemented.

FIG. 13.8

An example of a marinated pork tenderloin with peppercorn and garlic seasonings.

Courtesy of Smithfield Foods.

FROZEN CONVENIENCE PRODUCTS

These are valuable and palatable products for the busy "on-the-go consumer" who wants to spend less time on meat preparation. These frozen products can project a gourmet image, are attractively packaged, provide nutritional and preparation information, are microwaveable, and have a wide variety of choices. An example of Italian Style Meatballs is shown in Fig. 13.9. Tasty, frozen, convenient products prepared in a matter of minutes, such as chicken cordon bleu; beef stroganoff; and complete meals with beef, pork, veal, chicken, and turkey with potatoes, green beans, and so on, are examples of entrees that are available in retail stores.

CANNED MEATS

STERILIZATION PROCESSING OF CANNED MEAT

These products provide a preserved form of meat for an indefinite period of time under different environmental situations by sterilization. Sterilization is accomplished by heating a meat product in a hermetically sealed can to 250°F at 12–15 pounds/in.^{2} pressure. This is termed retort cooking and it kills all the anaerobic vegetative and spore-forming bacteria, especially *Clostridium botulinum*, a food-poisoning microorganism. An example of a shelf-stable ham is shown in Fig. 13.10. Unfortunately, some flavor changes can occur at retort-cooking temperatures. In general, however, they are very palatable products. Many soups and vegetables are also retort cooked. Canned hams continue to be a popular canned meat product as they have good flavor and are stable for storage on a shelf at retail grocery stores. Many other canned meat products such as SPAM, stews, Vienna-style wieners, chili, and soups are also available for consumer purchases in retail stores in cans that are processed under retort-style cooking.

FIG. 13.9

An example of a high-quality frozen meat product (Italian Style Meatballs) that is fully cooked and easy to prepare.

Courtesy, Smithfield Foods.

FIG. 13.10

An example of a canned ham that is shelf stable and processed by retort cooking, resulting in a sterilized product.

PASTEURIZED CANNED PRODUCTS

Pasteurization is a process in which canned products are heated to 155–165°F. At this temperature, all microorganisms may not be killed, the canned product does not have an indefinite shelf life, and the product must be refrigerated to maintain the shelf life. Pasteurized canned hams would be an example of a pasteurized product available in many retail outlets. It is recommended that they be consumed within 6 months after processing to receive optimum palatability (Fig. 13.11).

READY-TO-EAT (RTE) MEATS

The RTE meat products from beef, pork, lamb, and poultry meat were developed to meet the demands for consumer convenience, and they require no further preparation by the consumer before eating. Ready-to-eat meats consist of a large variety of processed products that must meet zero tolerance for the pathogen *Listeria monocytogenes*, high performance standards for the pathogens *Salmonella*, *Escherichia coli*, *C. botulinum*, and *Clostridium perfringens* and strict guidelines on cooking, cooling, packaging, and storage. Because of the high performance standards, this style of meat is safe and ready to eat. Some examples of RTE meats are cured meats; sliced, cooked, and boneless hams; sliced cooked loaves; snack sticks; jerky; barbecued meats (pulled pork or beef); poultry rolls and breasts; roast beef; and other delicatessen items (Fig. 13.12).

FIG. 13.11

An example of a pasteurized canned ham that must be kept under refrigeration for storage.

FIG. 13.12

An example of ready-to-eat (RTE) meat products.

Courtesy of Smithfield Foods.

CURED AND PROCESSED MEATS

Curing, or salting, of meat is recognized historically as one of the oldest forms of preservation, dating back to 3000 BC. Preservation was achieved by applying high concentrations of salt to the lean surface of meat. This caused dehydration to the point of lowering moisture to a level where bacterial growth and spoilage were prevented. It also provided a safe and enjoyable product for consumption. Today, however, the main purpose of curing is to impart an attractive flavor and color to cured meat.

Two forms of applying curing ingredients are dry cures and pickle or brine cures. Dry cures are used to a limited extent to produce country-style, or dry-cured, hams and bacon for a niche market. This is a process where the dry-cure mixture (sugar, salt, nitrate or nitrite) is hand rubbed in excess over the outside of the ham, and this process requires several applications. Hams will be ready to merchandize after several months of curing under environmental or controlled temperature conditions. Some examples of dry-cured hams are Country ham, Parma, Westphalia, Smithfield,

Aging of country style hams

Final product of country style ham

FIG. 13.13

An example of a dry-cured, country-style ham.

Courtesy, Dr. Gregg Rentfrow and Dr. Swendranath Suman, Department of Animal and Food Science, University of Kentucky.

and Prosciutto (Fig. 13.13). The famous Jinhua ham produced in China is another style of a dry ham (Fig. 13.14). For some people, this kind of ham is the ultimate in flavor and eating enjoyment, but for other people, it may be too salty.

By far the most commonly used modern method of curing is termed pickle or brine cure. In the pickle, salt is the primary and absolutely essential ingredient for acceptable flavor of cured meats. Pickle curing is accomplished by dissolving different salt (NaCl) concentrations in water to form various concentrations (degrees) of pickles. These different concentrations of pickles are measured by using a salimeter and expressed as degree of pickle. Pickle cures can vary from 50° to 85°, with 65° pickles being the most commonly used. These pickles result in an acceptable level of salt that is desired by consumers, and it represents about 2%–3% in hams and 2% in bacon. When compared with dry-cured products, these salt levels are low concentrations, but they can help to inhibit some microbial growth. Because of low salt

FIG. 13.14

An example of a Chinese Jinhua ham.

From Animal Frontiers magazine, October 2012, vol. 2, No. 4.

concentrations, mechanical refrigeration is required to preserve and extend the shelf life and provide safe products for the consumer. The other curing ingredients in the pickle include sugar, nitrite, and adjuncts such as ascorbates, phosphates, and lactate.

The most common commercial method of introducing a pickle or brine into a cut of meat is a low-pressure pump injection system. Using an injection system, it speeds up curing, causes more rapid and uniform distribution of curing ingredients, and provides an easy way to produce marinated or enhanced products. In a ham, the cure can be uniformly injected via the vascular system. This is called artery pumping and has limited use in the industry. Stitch pumping is another method used to incorporate the brine into whole muscle products. A needle with multiple holes is inserted into the muscle in many locations, and the brine is injected. The most effective commercial injection system used today is multineedle injection. A series of perforated needles are arranged in an orderly array to uniformly distribute the cure throughout the interior of cuts as they pass through a conveyor track. A variation of stitch pumping, termed multiple needle injection, is primarily the method of choice employed by the industry in curing of hams and bellies. Meat cuts should have an optimal internal temperature of 34–40°F, and the pickle or brine should not exceed 40°F to obtain the best results in cure retention. Hams are usually pumped to 110% of their green weight. Higher percentages are allowed, but these must be labeled as water added (Table 13.1). Bacons can vary from 105% to 110% pump. Bacon must be shrunk back, however, to the original green weight after processing. A typical automated

Table 13.1 Assignment of cooked ham product by name based on the calculated protein fat-free (PFF) values in the ham products[a,b]

Minimum PFF	Product name and qualifying statement
20.5	Ham or cooked ham
18.5	Ham with natural juices
17	Ham, water added
<17	Ham and water products with *X*% added as ingredients

[a]*PFF is calculated as [(% meat protein)/(100 − % fat)] × 100.*
[b]*Code of Federal Regulations, Title 9: Animals and Animal Products, Chapter III, Subchapter A Part 319, subpart D, 319.104, Cured pork products.*

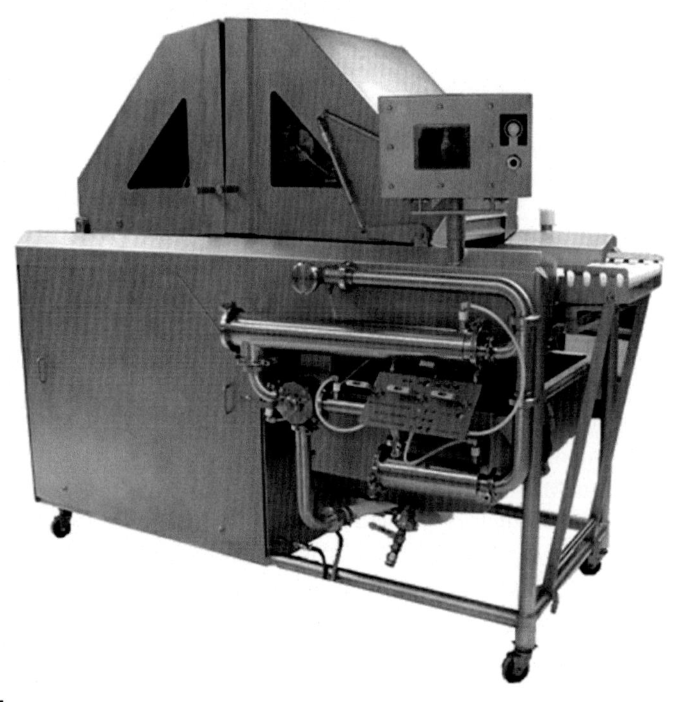

FIG. 13.15

An example of an injector system used to inject hams and other cuts of meat with a brine for curing meat.

Courtesy of Marel Meat Processing Inc., Des Moines, Iowa.

ham processing line can inject 150,000 pounds a day (Fig. 13.15). Massaging or tumbling of boneless hams acts to uniformly distribute cure, extract proteins for binding meat pieces, and obtain desired texture. Currently, in an effort to reduce curing times, large commercial processors move these products directly from the curing process to the thermal processing unit for cooking and smoking. After heating and smoking,

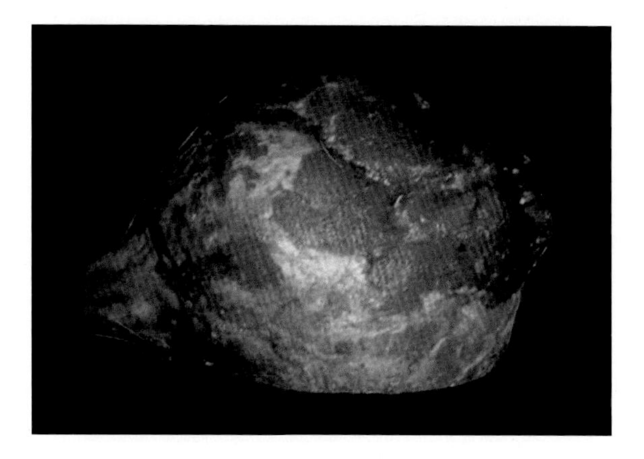

FIG. 13.16

An example of a bone-in cured and smoked ham.

products are refrigerated. Subsequently, cured products are packaged as whole, parts, or specialties such as slicing for retail sale. Hams can be bone-in, partially boneless (whole or part), and boneless. An example of a bone-in ham is in Fig. 13.16.

Bellies, hams, and picnic shoulders are the primary pork cuts cured and smoked. But some heavy boneless pork loins are cured and smoked and sliced into Canadian bacon, a popular topping for pizza. Another special product is the cured and smoked center section of the whole loin when fabricated into thick pork chops, commonly termed Windsor chops (Fig. 13.17). The most well-known cured beef is the cured boneless brisket, called corned beef. The addition of spices, garlic, and whole peppers are common ingredients used in the production of corned beef. With slow, moist

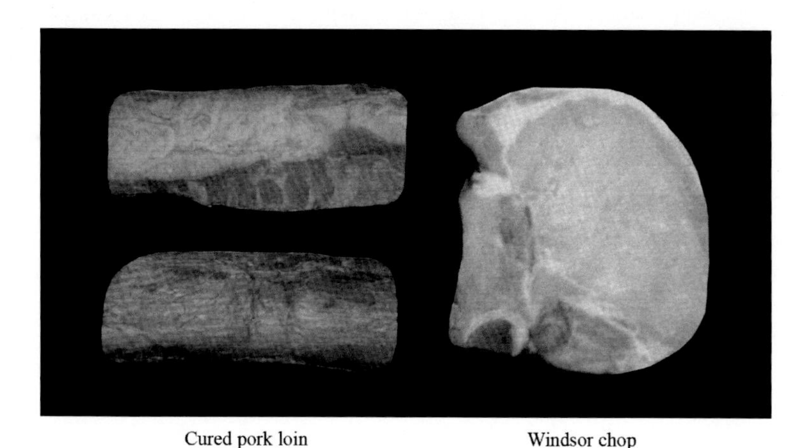

FIG. 13.17

An example of a Windsor-style cured pork loin and a center cut Windsor Chop obtained from the cured loin.

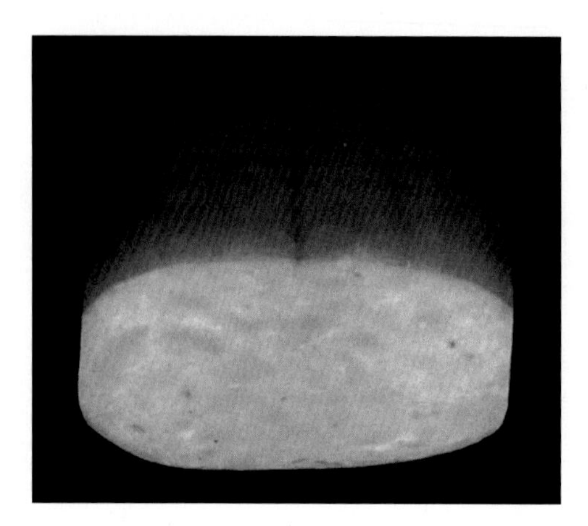

FIG. 13.18

An example of turkey ham formed from turkey thigh meat.

cooking, corned beef can be served as a roast-type dish or it can be cut across grain into thin slices for sandwiches such as the Reuben. Some muscles from the beef round are occasionally cured, smoked, and dried and sold as dried beef. The boneless beef plate can be cured and smoked and merchandized as beef bacon. The serratus ventralis muscle, a large muscle in the short ribs, is sometimes cured, smoked, and spiced for pastrami. Also, turkey thigh meat can be cured, formed, smoked, and cooked to become turkey ham (Fig. 13.18).

Factors influencing quality of cured meats are numerous, but two of these factors, high levels of sanitation and low refrigeration temperatures, are essential in the production of high-quality cured meats. The curing room should be maintained at refrigerated temperatures. Also, packaging, transporting, storage, and retail sales of the processed products all require proper sanitation and temperature control.

CURING INGREDIENTS

WATER

Added water plays a critical role in the quality and quantity of manufactured cured meat products. Added water is defined as the amount of water in excess of normal amounts in the unprocessed meat product. Water serves as the solvent for the curing ingredients, provides a moist and juicy meat product, compensates for moisture loss during thermal processing, and for the meat processor, results in more profitability.

The processor must give serious consideration to the quality and quantity of water used in making a pickle or brine for processed meat products. Insofar as quality is concerned, the practice of using soft water offers an advantage by dissolving phosphates more completely and resulting in a more desirably flavored and eye-appealing finished product. In addressing the quantity of water, added water must be tightly bound in the processed product to prevent unsightly purge (juice and drip loss from the cut) in a packaged product. Quantity of water added is regulated by FSIS, and regulations for water added and increased yields are based on the protein fat-free method (see Table 13.1). Furthermore, retail sale labels for packaged products must contain information about added water that results in added weight in cured and cooked red meat and poultry products.

SALT

Salt solubilizes the myofibrillar proteins. Soluble myofibrillar proteins make strong heat-set gels and thus improve the "bind" in processed products. Salt can also effectively lower the isoelectric point of proteins, and thus salt improves the water-holding capacity in meat systems (as long as the pH stays constant). As earlier indicated, salt (NaCl) is the most significant ingredient in the brine formula for the purposes of preservation, taste, and flavor enhancement. It must be of high purity, and the amount used depends on the production of either dry cured or modern-day injection of pickle or brine into the cut. Recently there has been a great deal of interest in reducing sodium in processed foods to improve the health of people. Eliminating or limiting the use of sodium chloride in processed meats is a challenge because the salt is so effective in preservation, slowing bacterial growth, increasing salt taste, enhancing flavor, and increasing water-holding capacity of the meat proteins.

NITRATE AND NITRITE

The chemistry and practical use of nitrate for preserving meat have a long and interesting history. Probably no compound has evoked so much controversy, stimulated as much scientific investigation, and had the impact on the cured meat industry as nitrate. Early meat processors used saltpeter (sodium or potassium nitrate) in their curing formulations, and it was likely a contaminant in salt. Research workers later discovered that nitrate was the effective ingredient, and furthermore, it was learned that a complex series of reactions involving nitrate was responsible for the formation of cured meat color. These chemical reactions are shown in Fig. 13.19. The initial chemical reaction is the reduction of nitrate and nitrite to nitric oxide, a gas. Then nitric oxide reacts with myoglobin and metmyoglobin to form the dark red cured meat pigment, nitrosomyoglobin. When the cured meat product goes through the thermal processing cycle, nitrosomyoglobin is chemically converted to the typical light pink cured meat pigment, nitrosohemochrome, a relatively stable pigment. The practical significance of the pigment, nitrosohemochrome, is that consumers associate it with the attractive color of cured meats. For cured meat color to remain attractive and salable, prevention of oxidation of the pigments to other undesirable color pigments is essential. The poor color pigments are characterized as oxidized porphyrins that have brown and green colors. The most effective way of preventing oxygen from causing discoloration and oxidative rancidity (off-flavors) in cured meat products is to package them in vacuum packages. Other variables negatively affecting cured meat color

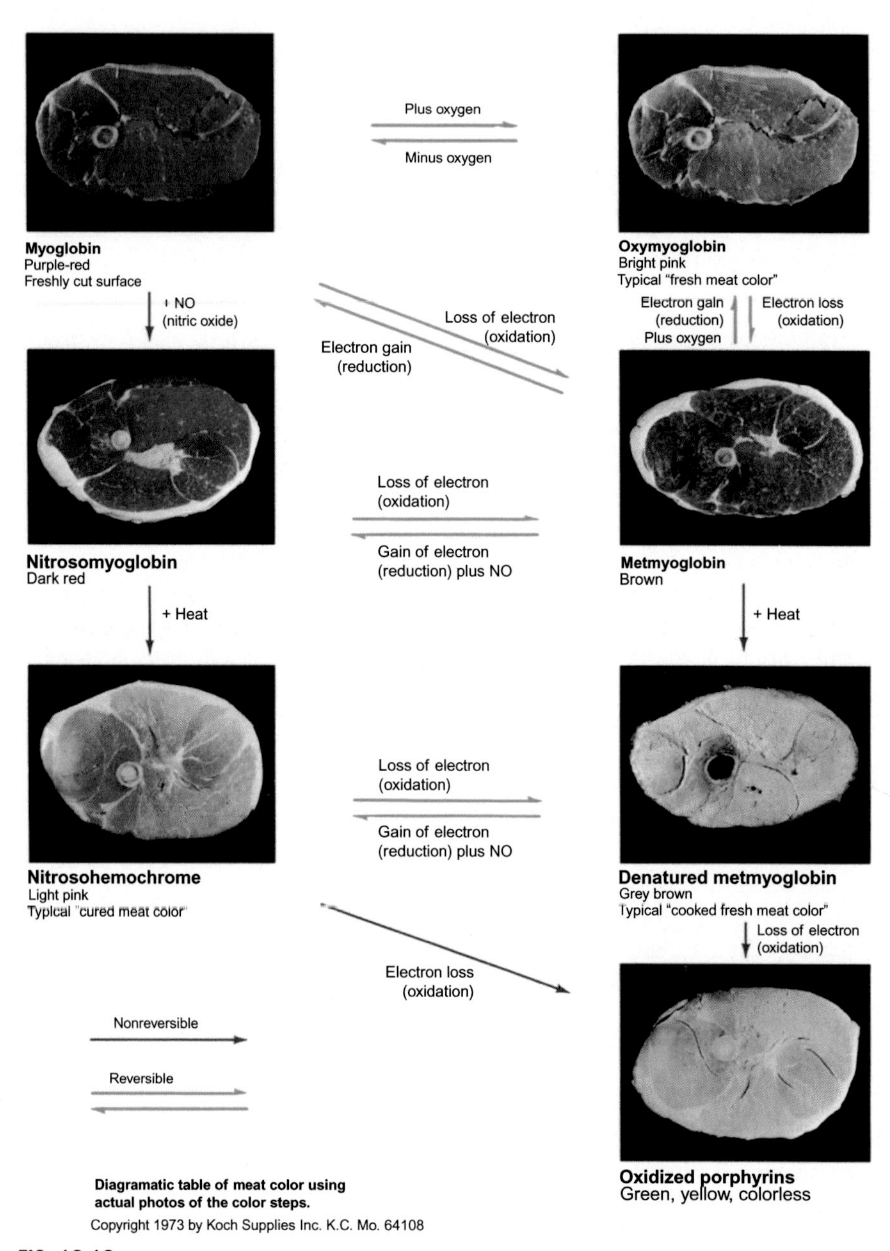

FIG. 13.19

The relationship between chemical reactions and meat color in cured hams is shown in this figure.

From Ultra Source, LLC, Kansas City, MO (formerly Koch Supplies, Inc.).

are improper retail-case display lighting, insufficient nitrite in the cure, uncontrolled bacterial growth, a poor state of cleanliness, and improper application of storage and retail case temperatures.

Today nitrate cannot be used with the exception of the production of dry-cured hams and some dry sausage products that require curing for extended periods of time or for specialized products. The reason for not being able to use nitrate in all other cured meats is that it is difficult to control the amount of residual nitrite in the product when nitrate was used.

The use of nitrate or a mixed nitrate-nitrite cure is not common because variable residual nitrate and nitrite concentrations can result in production of nitrosamines. Nitrosamines are carcinogenic compounds. Hence, curing mixtures today contain sodium nitrite, or other sources of organic nitrite such as celery powder in the curing formulation. The use of organic or natural cures is popular among some consumers, although they may not be as effective as inorganic nitrite in preventing microbiological contamination of meat products. Naturally cured meats and poultry products are foods that are rapidly growing in consumer popularity. They must be labeled, however, as natural meat products. The label would read as follows: natural cure, no nitrates or nitrite added except for naturally occurring nitrates found in natural plants such as celery. The label would also indicate that the product is not preserved and should be refrigerated below 40°F at all times.

Nitrite serves several beneficial purposes in cured meat, and as yet, no other compound has been found to replace nitrite. Nitrite fixes the cured meat color, stabilizes and adds flavor, and from a significant public health standpoint, inhibits the formation of *C. botulinum* spores. The desired flavor of cured meats is thought to come by nitrite acting as an antioxidant and preventing the formation of oxidative products. Obviously, nitrite is very important in the salability of cured meats, especially to the pork industry as over half of the pork carcass is merchandized in some form of cured products. Another positive feature of nitrates, nitrites, and nitric oxides, recognized recently, is the beneficial physiological effects on cardiovascular and immune function for improved health of humans.

Nitrates from vegetable consumption and nitric oxide from cured meat seem to play a role in controlling blood pressure, immune response, wound repair, and neurological functions. Nitric oxide from the reduction of nitrite may control human blood flow in heart muscle and prevent certain kinds of cardiovascular diseases including hypertension, atherosclerosis, and stroke. Federal regulations limit the use of sodium nitrate and nitrite because these compounds taken in excess can also be highly toxic to humans. According to regulations, sodium nitrite should not exceed 200 ppm in the product after curing and processing. The maximum for bacon is 120 ppm. The maximum amounts of sodium nitrite and/or potassium nitrite that may be used in curing are as follows:

- 2 pounds in 100 gal of pickle,
- 1 oz. for each 100 pounds of meat in dry salt, dry cure, or box cure, and
- 1/4 oz. in 100 pounds of chopped meat and/or meat by-products.

SUGAR

The main functions of sugar and other sweeteners are for providing a sweet flavor to the product and reducing the harshness of salt. The most common sugar used is sucrose. Corn syrup and corn syrup solids are also used as substitutes in curing pickles, and they are usually lower in cost than sucrose. Ham pumping pickles vary in content of sugar, but 50 pounds per 100 gal seem to be the most common amount, and bacon pickles frequently use 75 pounds of sugar per 100 gal of pickle.

ASCORBIC ACID AND ERYTHORBIC ACID

Ascorbic acid and erythorbic acid are considered curing adjuncts. They are not necessary but helpful in the production of a higher quality cured product. Ascorbic acid and erythorbic acid and their salts are used to hasten the development of the cured meat color by acting as reducing agents to speed up the breakdown of nitrite to nitric oxide and keep myoglobin in a reduced form so that it will react with nitric oxide. Ascorbic acid and erythorbic acid must always be added last when mixing the brine to prevent the untimely reaction of nitrite to nitric oxide (a gas), which would not be available to properly cure the product. These acids and their salts (sodium ascorbate and sodium erythorbate) also act as cured meat color stabilizers and antioxidants. Because these agents hasten the conversion of nitrite to nitric oxide, there is less residual nitrite in the finished product. This also means that ascorbates reduce or inhibit the formation of nitrosamines. Federal regulations permit addition of 0.75 oz. of ascorbic acid or erythorbic acid or 0.875 oz. sodium ascorbate or erythorbate per 100 pounds of chopped meat, or the addition of 75 oz. of ascorbic acid or 87.5 oz. of sodium ascorbate or erythorbate to 100 gal of pickle for curing primal cuts.

ALKALINE PHOSPHATES

Alkaline phosphates have found strong use in the meat industry because they increase water-holding capacity and, therefore increase the yield and dollar value for cured meat products. Sodium tripolyphosphate, sodium hexametaphosphate, sodium pyrophosphate, and disodium phosphate have all been approved for production of cured pork and beef cuts. Legal limits for residual phosphates are restricted to not >0.5% in the finished products. Phosphates act by adjusting the pH to the alkaline side of the pH (equal electrical charges) and as a result increase the water-binding capacity. They also chelate (bind) bivalent metal ions, for example, calcium, and increase ionic strength, which also allows for an increased water-binding capacity. Phosphates at not >0.05% must always be added first to the water when making a brine or pickle to ensure that they are totally dissolved before adding any other ingredients.

ANTIMICROBIALS

Sodium or potassium lactate is the salt of lactic acid. Lactate has a wide variety of applications. One of its most important properties, particularly from a food safety perspective, is its effectiveness as an antimicrobial agent. It functions primarily by reducing the water activity of foods. Sodium lactate controls both aerobic and anaerobic bacteria. When solutions of lactate are sprayed on beef, pork, and poultry carcasses, pathogens are reduced. The growth of two major pathogens, *C. botulinum* and *Listeria monocytogenes*, is inhibited by sodium lactate. From a meat-palatability standpoint, meat injected with solutions containing lactate is more tender, juicy, and flavorful. Flavor in both fresh and processed meats is improved because of the antioxidant property of lactate. Moreover, retail shelf life of the meat that contains lactate is longer. Lactate may also have an important role in the development and stability of cured meat color by producing more reduced nitrite. Many other antimicrobials have been introduced and approved for use in processed meat in recent years.

SMOKING AND COOKING

Historically, smoking meat was a way to preserve it, usually in combination with salting. Today, the purposes of smoking are to impart a characteristic flavor and aroma, give a uniform mahogany-colored surface, act as antioxidants, serve as bacteriostatic agents, and coagulate proteins to form a thin skin on the outside surface of the sausage products. The thin skin formation is important in the manufacture of skinless wieners (Fig. 13.20).

Natural wood smoke is commonly generated from hardwoods such as oak, hickory, apple, cherry, and mesquite. For smoke to be properly absorbed on the surface of the meat, the meat surface must be slightly moist before smoke application. A slightly moist surface will result in a uniform mahogany color. If the surface is too wet, a streaked appearance will result, and if the surface is too dry, the smoke will adhere poorly and the color will be pale. Good smoke color can be obtained by careful use of a drying cycle consisting of a low temperature (~100°F), low relative humidity (<45%), and rapid air movement.

The best-quality smoke is produced at a combustion temperature of 650–750°F, but lower temperatures (650°F) are used because carcinogenic substances (benzo[a] pyrenes) can be generated at 750°F. Although smoke at the point of generation exists

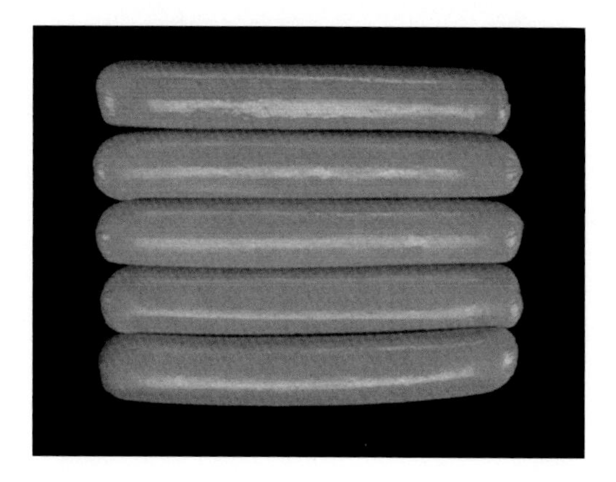

FIG. 13.20

An example of skinless wieners.

in a gaseous state, it rapidly partitions into a vapor and particle state. The vapor phase contains the more volatile components, and it is largely responsible for the characteristic flavor and aroma of smoke. Over 200 different compounds have been isolated from wood smoke. However, components of smoke such as phenols, acids, alcohols, and carbonyl compounds have significant effects on smoked meat quality. The phenolic compounds are largely responsible for the functions of smoke. Another option for the meat-processing industry is the use of liquid smoke preparations, an aqueous solution of natural hardwood smoke, by spraying or atomization of the solution onto the surface of the meat product just before cooking. It can also be added directly to the meat mixture. Liquid smoke, because it is cleaner, requires less expensive equipment, and does not contain carcinogenic compounds, is replacing natural wood smoke in many processing operations.

For the production of high-quality smoked and cured meats, the meat products are heated (cooked) in a smokehouse or thermal processing unit (Fig. 13.21). A thermal processing unit, such as a smokehouse, is a highly sophisticated piece of equipment consisting of a programmable computer to control temperature, humidity, smoke, and relative humidity. The temperature is slowly elevated to begin the cook cycle, smoke will be applied, and further cooking will take place with added humidity to prevent surface or case hardening. After the required internal temperature is reached, the cycle is completed by applying a cold shower, and the cured product is transferred to a chilled storage room. Proper use of the thermal processing unit will produce highly attractive and palatable products designed for consumer satisfaction.

FIG. 13.21

An example of a smokehouse or processing oven for processing meat and meat products.

DRYING

Drying is perhaps the earliest method used to preserve meat, and it remains very effective today. Other natural or physical methods of drying meat reduce intrinsic moisture and thus prevent the growth of microorganisms. A popular form of dried meat today is jerky. It is usually sold as beef jerky, but it can be made and sold from the muscle of game, lamb and mutton, and other meat sources. Jerky is a dried product manufactured by using a combination of curing, smoking, and drying. For a jerky product, the water-to-protein ratio must be 0.75–1, meaning it has a range of 30%–40% moisture in the jerky and an α_w of 0.55–0.70 (low moisture) preventing bacterial growth, and it does not require refrigeration. The main function of smoking and curing jerky is for flavor. Because of its popularity, dried meats such as jerky products are widely available for sale, particularly at convenience stores.

PACKAGING

Packaging of cured meats for optimal eating quality and salability requires vacuum packaging equipment and oxygen-impermeable packaging materials. Because oxygen is not required for cured meat color formulation, air must be prevented from entering the package to prevent discoloration and development of objectionable rancid flavors. Hence, vacuum packaging of cured meat cuts is essential to optimize color, flavor, and extend shelf life. Refer to Chapter 15 for examples of different packaging styles and systems for meat.

SUMMARY

Processed meats are manufactured from fat and muscle of wholesale cuts, trimmings from carcasses, and some nonmuscle cuts such as liver. Processed meats can be divided into fresh processed meat and cured and smoked processed meat. Examples of fresh processed meat are ground beef, mechanically separated meat (MSM) from bone, mechanical and enzyme tenderization of steaks, moisture and marination of fresh pork products, and frozen meat items. Examples of other processed meat products are canned ham, canned SPAM, canned soups with ham, Vienna-style wieners, and stews. Ready-to-eat meats (RTE) can also be obtained in sliced cooked loaves, snack sticks, jerky, roast beef, and other delicatessen items.

Cured and processed meat products make up a large number of the processed meat items sold in the United States and Europe. Curing and salting meat is one of the oldest forms of meat preservation. Earlier than 3000 BC, the salting of meat for preservation was practiced. Most curing procedures use a salt-based brine or pickle for manufacturing the cured meat products. These curing ingredients have a major influence on the flavor of cured meat. The most common commercial method for introducing the brine into a cut of meat is by a low-pressure pump injection system. These methods are used for curing and processing hams and bacon as well as Windsor Chops and Canadian bacon. Other examples of cured and processed meat items are dried beef, cured beef brisket called corned beef, and turkey ham. Another important ingredient in the brine mixture is sodium nitrite. It is responsible for the cured meat color, adds flavor, and can help prevent the development of spores from microorganisms such as *C. botulinum*. Most cured meat products are also heated or cooked in a smokehouse to enhance the flavor and surface color of the meat. To maintain good color of the cured meat products, packaging the meat with oxygen-impermeable packaging materials is essential.

QUESTIONS FOR STUDY AND DISCUSSION TOPICS

1. Define processed meat and meat products.
2. Provide examples of fresh processed meat and meat products.
3. Define mechanically separated meat (MSM).
4. What type of beef cuts are used for mechanical and enzyme tenderization? Provide examples of the procedures used for these two tenderization methods.
5. Describe two different methods that are used in the commercial industry for canned meat.
6. Describe two methods for applying the curing ingredients to meat.
7. What are the primary wholesale pork cuts used for the curing and smoking process?
8. When curing ingredients are evaluated for function, what is the most important ingredient? Describe its function in the curing process.
9. What is the major function of phosphates in the curing process?
10. Natural wood smoke is commonly generated from what type of wood?

CHAPTER

Sausage processing and production

14

INTRODUCTION

Sausage can be defined as a food prepared from comminuted (particle size reduction) and seasoned meat formed into various symmetrical shapes in which the products differ primarily in the variety of spices used and the different processing methods applied. There are several hundred varieties of sausage available today because our European ancestors were innovative and creative "wurstmachers," and this tradition continues today.

Sausage is one of the oldest forms of processed food, and it originated, in part, as a means of preserving meat. The word "sausage" is derived from the Latin word, *salsus*, meaning preserved or literally salted. Sausage was made and eaten by the Babylonians and ancient Chinese some 1500 years BC. Salami was frequently mentioned by the Greeks about 449 BC, and the American Indian made a sausage product called pemmican. It was made from dried meat that was pounded into a powder and then added to melted fat with various berries. Certain sausage products became especially popular, and they took their name from the city or village in which they originated, for example, frankfurters—Frankfurt, Germany; bologna—Bologna, Italy. Out of necessity, dry, low-moisture sausages were developed to preserve them in the warm climate of Southern Europe. They were preserved in this manner because there was no refrigeration, whereas fresh or semifresh products were developed in the cooler climates of Northern Europe. Therefore sausage processing has developed from a simple process of salting and drying meats to the many complex seasoning and processing procedures of today. Consequently, consumers have hundreds of sausages to select from. Indeed, processed meat manufacturers have developed creative and innovative ways to make tasty, convenient, healthful, and value-added sausage products. This chapter will provide an overview of the common sausages marketed in the United States.

SAUSAGE INGREDIENTS

RAW MEAT MATERIALS

The proper selection of raw materials, primarily animal tissues, is fundamental to the production of uniform, high quality, safe sausage products. In other words, a sausage product will be no better than the raw meat ingredients used to manufacture it.

The Science of Animal Growth and Meat Technology. https://doi.org/10.1016/B978-0-12-815277-5.00014-7

Table 14.1 Examples of different meats used in sausage manufacture (classified according to binding[a] qualities)

Items	Examples
Meats having the best binding qualities:	Bull beef
	Boneless cow meat
	Beef chucks
Meats having fairly good binding qualities:	Beef head meat
	Beef cheeks
	Boneless veal
	Calf head meat
	Pork trimmings, extra lean
	Pork trimmings, lean
	Pork head meat
	Pork cheeks
Meats having poor binding qualities:	Beef hearts
	Beef weasand meat
	Beef giblets
	Beef tongue trimmings
	Regular pork trimmings
	Pork hearts
	Pork jowls
	Pork ham fat
	Sheep cheeks
	Sheep hearts
The following meats, although nutritious, have practically no binding qualities at all and are used as fillers in the interest of economy. The use of these ingredients should be limited to less than 25% of the meat formula because of their high connective tissue content.	Ox lips
	Beef tripe
	Pork snouts
	Pork lips
	Pork tripe

[a]Bind is defined as the capacity to attract and retain water and encapsulate fat.

In Table 14.1, a wide variety of different raw meat products used in the manufacture of sausage products are presented. These raw materials are derived from the skeletal and nonskeletal parts (variety meats) of beef, pigs, sheep, and calves. In addition to these tissues, chicken and turkey skeletal muscle meats are also used. Poultry meat is a significant ingredient in the manufacture of sausage products because it is a less expensive ingredient than red meats. Mechanically separated meat from beef, lamb, pork, chicken, turkey, and minced fish are raw materials used in the manufacture of some sausage products. The most valuable raw meat material ingredients for sausage production are lean skeletal meats obtained from bull and cow carcasses. This beef is especially recognized for its high lean-to-fat ratio, lean color, and superior water-binding capacity. Pork and beef trimmings provide most of the added fat in a sausage formulation. Some nonskeletal muscle meats such as hearts, tongues, and tripe are used sparingly as filler meats in some least-cost formulations. When used, all filler meats must be listed individually on the package ingredient statement. Other nonskeletal muscle meats, however, are used to make high quality sausage items such as liver sausage and blood and tongue sausage. Because of the variation in animal tissues, certain standards of freshness, bacterial contamination, binding properties, lean-to-fat ratio, composition, lean color, collagen content, bone and bone fragments, and "gristle"-free meat must be met to achieve high quality sausage products.

The characteristics of meat associated with binding ability are particularly essential in the production of high quality sausages. Naturally, meat has a wide variation in its ability to bind water and hold lean and fat together. The best water-retention and fat-emulsification properties of meat during processing are from the proteins of lean skeletal muscle tissues.

MYOFIBRILLAR PROTEINS

The myofibrillar proteins are the class of meat proteins responsible for binding water and encapsulating fat. For example, the addition of salt to the lean meat portion of the sausage formulation to solubilize the myofibrillar protein is the most effective way of binding added water and fat in the sausage product. Once solubilized, the myofibrillar proteins are able to form a "gel" when heated. This gel matrix is the explanation for the change in texture of cooked salted meat (e.g., from frankfurter batter to a frankfurter that has texture and bind).

SARCOPLASMIC PROTEINS

Another class of meat proteins is called sarcoplasmic proteins. They are water soluble, and the muscle pigment, myoglobin, is located in the sarcoplasmic protein fraction. Myoglobin is responsible for the color patterns in meat. For example, beef has the most myoglobin and has the darkest color. Pork has less myoglobin and has a lighter color than beef. Poultry breast and wing meat has the least myoglobin and is lighter in color than pork. The darker poultry meat from legs has more myoglobin and is darker than breast meat. Therefore the sarcoplasmic protein traits will influence the color of processed meat. The sarcoplasmic proteins, however, have very limited binding capacity when compared with the myofibrillar proteins. More detailed information on sarcoplasmic proteins can be found in Chapter 9 and Fig. 9.6.

STROMAL PROTEIN

Collagen is an example of stromal protein in muscle. This muscle connective tissue protein is capable of binding water, but collagen does not have good binding traits. Collagen-type meats such as beef shanks should be limited in the manufacture of an emulsion-type sausage to obtain a stable batter.

FAT

In addition to the muscle component for sausage, fat from beef and pork trimmings is added to adjust to the desired fat content in the sausage formulation. Fat contributes to the texture, juiciness, taste, flavor, and final price of the sausage. Fat is a major ingredient in sausage production. Today, however, the trend is to decrease fat content to meet the consumer demand for low-fat or fat-free sausages.

NONMEAT INGREDIENTS

Sausage makers have a wide variety of nonmeat products available to improve the functional properties of the sausage, enhance flavor, increase cooking yields, protect against microbiological contamination, extend shelf life, and increase profitability.

WATER

Added water in the form of liquid, ice, or both is a significant nonmeat ingredient in a sausage formulation for both processor and consumer satisfaction. For the processor, it is a critical ingredient for dissolving and dispersing curing ingredients and controlling temperature during processing. Moreover, water quality is important because using soft water eliminates metals that are harmful to the processed meat color. For consumer appeal, water affects color, appearance, and palatability (flavor, tenderness, juiciness). Insofar as water quantity is concerned, water in fresh sausages and luncheon meats is limited to 3% by government regulations. For cooked and smoked sausage, the federal regulation is 10% added water. For emulsion-type sausages, government regulations have been modified to allow water to replace fat. Because of the demand for low-fat or fat-free emulsion-type products, added water is no longer limited to 10%, but rather the sum of fat and added water cannot exceed 40%. Maximum allowable fat content is 30% in emulsion products. It is essential that added water be bound and entrapped within the sausage structure to yield high quality sausage products.

SALT

Sodium chloride (NaCl) is used in most processed meat formulations. It serves two very essential functions. It is absolutely essential in producing a desirable flavor in sausage products. Without the addition of salt, the product is unacceptable. Second, salt is essential for the solubilization of myofibrillar proteins. These soluble proteins are needed to bind water and fat to provide a high quality sausage. Most sausages contain about 1.5%–2.5% added salt. Moreover, salt must be of high purity because impurities may act as prooxidants in the development of oxidative rancidity resulting in off-flavors. Some manufacturers use a NaCl substitute in sausage formulations such as potassium chloride or use a combination of sodium and potassium chloride salt. Potassium chloride has a bitter taste so it is not used above 50% of the total salt content. The reason for using a salt (NaCl) substitute or using less sodium chloride is that sodium is considered the culprit in human hypertension. Therefore some consumers want an option to purchase sausages with a lower content of sodium chloride or having a sodium chloride substitute such as potassium chloride.

NITRATES AND NITRITES FOR MEAT COLOR AND FLAVOR

Early meat processors recognized that saltpeter was the ingredient that caused a good color and flavor to cured meat. Subsequently, scientists found this ingredient to be nitrate. Extensive research eventually showed that nitrite (NO_2) not nitrate (NO_3) was the substance responsible for the cured meat color and cured meat flavor.

The use of nitrate today is limited to the production of dry-cured hams and dry sausages. Currently, sodium nitrite is added directly to the sausage formulation. In the formulation, nitrite is reduced to nitric oxide (a gas), and through a series of complex chemical reactions, the cured meat color is formed (see Fig. 13.19). In whole-muscle cuts such as ham, the cured meat color development is rather straightforward. In sausage manufacture, however, the meat is often chopped, and this incorporates oxygen from the air into the meat. Oxygen then combines with myoglobin (meat pigment) to form a bright red meat color, oxymyoglobin. This red color is the result of an oxygenation chemical reaction. With further exposure of oxygen to the meat pigment, oxymyoglobin is oxidized and the meat color turns brown, an undesirable color. This color is termed "metmyoglobin." Both myoglobin and metmyoglobin can combine with nitric oxide to form the pigment nitrosomyoglobin. Nitrosomyoglobin, upon heating, is converted to the stable cured meat color pigment called nitrosohemochrome (see Fig. 13.19). This is the reddish-pink color of cured hams that is desired by consumers. The same cured color is in frankfurters as well as all meat products cured with nitrite. To maintain and extend the cured meat color when in a package, oxygen must be excluded from the package. This is achieved by vacuum packaging. The use of nitrate and nitrite is both highly regulated because, when consumed in excess, they are toxic to humans. FSIS regulations allow the safe use of sodium or potassium nitrite to not exceed 1/4 oz. (156 ppm) per 100 pounds of meat in sausage products. Another benefit of nitrite is its excellent antioxidant and antimicrobiological traits. These also contribute to the cured meat flavor.

ASCORBATES AND ERYTHORBATES

Ascorbates and erythorbates are reducing agents involved in cure acceleration by reacting with nitrite to produce nitric oxide. They can also reduce metmyoglobin to myoglobin. These reactions can help stabilize the red color of the cured meat pigment, nitrosohemochrome. From a food safety perspective, ascorbates and erythorbates inhibit the formation of nitrosamines (cancer-causing compounds) in a sausage product. These reducing agents are regulated by USDA-FSIS at 3/4 oz. per 100 pounds of meat (acid form) and 7/8 oz. per 100 pounds of meat (salt form).

SUGARS

A wide variety of sugars are available to the sausage maker. Sucrose, dextrose, corn syrup or corn syrup solids, lactose (in nonfat dry milk), and sorbitol are examples. The primary function of sugar in a sausage formulation is to serve as a sweetener. In fermented sausage, however, dextrose is most often used as the substrate for production of lactic acid by the metabolic action of lactic acid bacteria. The lactic acid produced lowers the pH and contributes to the characteristic "tangy taste" of summer sausage and helps with the safety of the sausage during storage because it prevents bacterial growth.

ALKALINE PHOSPHATES

A host of phosphates are available to be used in an amount not to exceed 0.5% of the finished sausage product. Alkaline phosphates improve the quality of sausage products by improving water-holding capacity and protein solubility, acting as antioxidants, and serving to stabilize the color and flavor of the finished sausage product.

ANTIOXIDANTS

Antioxidants are approved chemicals added to the fresh and dry sausage formulation to inhibit oxidation and prevent the development of "off-flavors" such as rancidity. They are allowed only in fresh and dry sausage. Fat in the sausage contains fatty acids, and the fatty acids are subjected to oxygen during the manufacture process. The fatty acids become vulnerable to oxidation reactions, and rancidity can develop. Other variables in the sausage production process can also stimulate oxidation in the sausage. Some examples are salt impurities, light, storage at higher temperatures, and certain metals. The antioxidants will eliminate or greatly reduce the oxidation process and reduce rancidity development and off-flavors. Some popular antioxidants are butylated hydroxyanisole (BHA), butylated hydroxytoluene (BHT), and propyl gallate. Government regulations allow 0.01% antioxidant of the fat content of the fresh sausage for only one antioxidant and 0.02% for any two or more in combination in fresh sausage. When used in dry sausages, 0.003% of the meat weight is allowed for any one antioxidant and 0.006% for the use of two or more antioxidants in combination.

OTHER NONMEAT INGREDIENTS

Other nonmeat ingredients are used to extend the shelf life by inhibiting microbial growth; improve the flavor; provide certain functional properties, for example, increased water-holding capacity; and decrease the cost of sausage products. Some of these are mold inhibitors such as potassium sorbate, some are cure accelerators such as glucono delta lactone, and some are flavor enhancers such as monosodium glutamate (MSG). Some nonmeat ingredients may be or contain allergens. Allergens are specific proteins derived from dairy products, soybeans, eggs, fish, and nuts. Therefore it is very important to list any allergens on the label of sausage products to provide a warning to consumers who may have allergic reactions to these allergens.

Meat extenders and binders are a large class of nonmeat ingredients, which include soy flour, soy protein isolates, nonfat dry milk, dried whey, reduced lactose whey, whey protein concentrate, sodium caseinate, wheat gluten, cereal flours, and tapioca dextrin. These products are limited to 3.5% of the finished sausage product with the exception of soy protein isolates. Soy protein isolate can be used up to 2.5%.

SEASONINGS, SPICES, AND FLAVORINGS

Many seasonings, spices, and flavorings serve to improve sausage flavor in many unique sausage products. This results in an international flair because many of them are produced from plants grown in many different parts of the world.

Seasonings can include a broad array of ingredients. Some examples include spices, herbs, and even vegetables. Spices are aromatic substances of vegetable origin with the active part residing in their volatile oils, many of which are grown on tropical islands. Some examples used in sausage manufacture are nutmeg, mace, garlic, sage, paprika, allspice, cinnamon, clove, ginger, and pepper. Black pepper in whole, ground, and cracked forms is the most popular spice in sausage manufacture. Most sausages contain pepper, especially frankfurters, bologna, pork sausage, summer sausage, and salami. Aromatic seeds include celery, caraway, and mustard. Sage, bay leaves, and thyme are condimental herbs, and onion and garlic are classified as condimental vegetables. Spices are used either whole or in one of the following processed forms: (a) ground, (b) essential oils, or (c) oleoresins. Most are used in their processed forms.

Flavorings are the product of spice extractives derived mainly from fruits, vegetables, herbs, and roots. Spice extractives are in the forms of essential oils and oleoresins. Essential oils are defined as volatile oils removed from plants while oleoresins are viscous, resinous materials obtained by extraction of ground spices with volatile solvents. Essential oils and oleoresins are available in either liquid or dry forms, and they must be labeled as flavorings.

SAUSAGE CLASSIFICATION

A clear and concise classification of sausages is complicated because of the wide variety of sausage types, formulations, and processes. The most reasonable and simple approach to classification may be based on the method of processing, namely, fresh, cooked, cooked and smoked (coarse ground and emulsion), dry and semidry, and cooked specialties such as loaves and luncheon meats. These sausage types will be discussed, and the practical as well as the scientific base for the manufacture of these sausages will be presented.

FRESH SAUSAGE AND THEIR MANUFACTURE

Fresh (not cooked or cured) sausages are made from selected cuts of fresh meat, mostly boneless pork and pork trimmings. Some fresh sausages are also made from beef and turkey meat. The USDA allows a lot of fat to be used in the manufacture of fresh sausages. An example would be fresh pork sausage. Fifty percent fat is allowed for the manufacture of fresh pork sausage, but the majority of the manufacturers will use 35% fat or less because the consumers are buying fresh pork sausage with less fat because it is better for health, results in less shrink when cooked, and has a less greasy texture or "mouthfeel." Some examples for marketing fresh pork sausage are in links (Fig. 14.1) or patties and may be sold as country-style, breakfast style, and "whole-hog" pork sausage.

MANUFACTURE OF FRESH PORK SAUSAGE

When fresh pork sausage is manufactured, fresh pork cuts and trimmings are chilled to 28–32°F to minimize fat smearing during the grinding process. When smearing is prevented, the ground pork has a bright red lean color, white fat, and desirable particle size. After grinding, the pork is seasoned with salt, pepper, and other seasonings and thoroughly mixed. Some pork sausages may have additional spices added such as sage, red pepper, ginger, and thyme to alter the traditional flavor of fresh pork sausage for special markets. After proper mixing, the sausage is stuffed into casings and stored in refrigerated areas.

FIG. 14.1

An example of fresh pork sausage links.

A special protocol is required for the manufacture of whole-hog sausage. Whole-hog sausage is made by boning (removing muscle and fat) from the prerigor pork carcass shortly after slaughter and before the carcass is chilled. This process is called "hot processing" of the pork carcass. Prerigor pork has a higher pH because the pH decline is interrupted with chilling and salting. A major advantage of hot processing is the ability for the myofibrillar protein in the muscle to bind more water resulting in a sausage that is more juicy. Prerigor pork is a common ingredient for many types of fresh sausage. The meat is processed before the pH declines and before the rigor bonds can form. Therefore the water-holding capacity, fat-binding ability, and color are superior to postrigor pork of the same cuts.

Bratwurst (Fig. 14.2) is another type of fresh pork sausage. Bratwurst is a German word for pork sausage. It is manufactured from ground pork trimmings, seasoned with salt, black pepper, nutmeg, mace, and coriander. After the spices are mixed with the ground pork, the mixture is stuffed into collagen casings. Bratwurst has become very popular at picnics and tailgate parties.

Another sausage that falls into the fresh category is thuringer, a breakfast-type sausage that contains veal, pork, and sometimes a small amount of beef, milk, chives, eggs, and parsley in the formulation. Bockwurst is another example. It has primary veal and some pork. It is sold in link casings that are the size of frankfurters. The fresh sausages are often sold in the frozen form to extend shelf life.

UNCOOKED SMOKED SAUSAGES AND THEIR MANUFACTURE

Uncooked smoked sausages are products made from cured or uncured meat; ground and mixed with spices, salt, or other nonmeat items; and stuffed into casings. The product is then smoked and refrigerated for storage. Some well-known varieties are

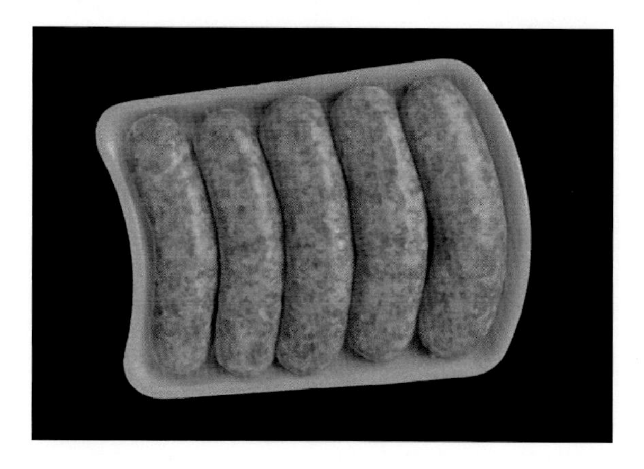

FIG. 14.2

Bratwurst, an example of fresh, not cooked or cured sausage.

smoked country-style pork sausage and Mettwurst, which is composed of approximately 60%–70% cured beef and 30%–40% cured pork. The major spices are pepper and coriander. Because these sausages are not cooked before they are placed in the retail market, they must be fully cooked before they are served or consumed.

COOKED AND SMOKED SAUSAGE AND THEIR MANUFACTURE

There are many cooked and smoked sausages manufactured from fresh cuts of meat. This is a very popular class of sausage because this class includes emulsion-type and coarse-ground sausages such as ring bologna (Fig. 14.3). Fig. 14.4 is an example of an emulsion-type product from a loaf such as olive loaf.

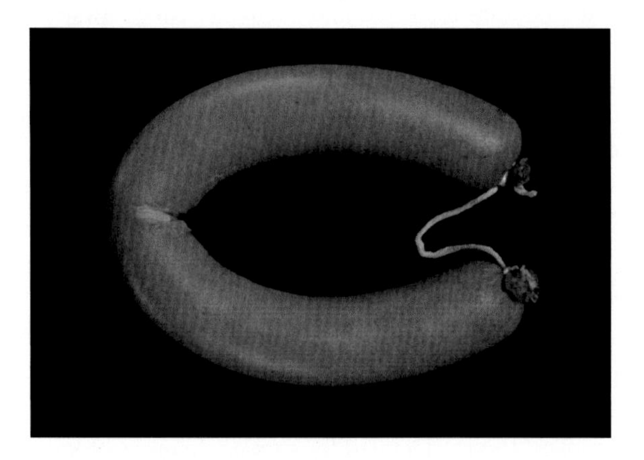

FIG. 14.3

An example of bologna as a smoked and cooked sausage.

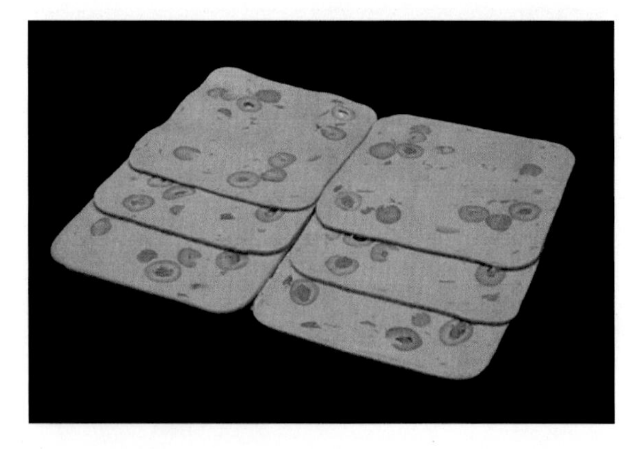

FIG. 14.4

An example of olive loaf, which is often used as a luncheon meat.

BATTER EMULSION CHARACTERISTICS AND THE CHOPPING PROCESS

A true emulsion is the dispersion of one immiscible liquid in another. For example, oil and water. Oil and water do not mix. To allow them to mix, an emulsifier is added such as lecithin. The emulsifiers prevent the separation of water and oil. The meat emulsion, however, differs from a true emulsion as it is a dispersion of fat particles in a matrix of solubilized proteins, water, and nonmeat ingredients. Therefore *a meat emulsion is more of a batter than an emulsion*. An emulsifier is a substance that is involved in forming and stabilizing the batter-emulsion matrix. Primary emulsifiers are the salt-soluble (myofibrillar) proteins. These salt-soluble proteins must be in solution to form a stable batter-emulsion matrix. The interaction of salt with the myofibrillar protein results in a soluble protein. The chopping process further releases the soluble myofibrillar protein to interact with the fat particles during chopping. As a result, the solubilized proteins and water of the sausage mixture form a matrix that encapsulates the fat particles, and the sausage batter is formed. An example of the matrix that encapsulates the fat particles with the soluble myofibrillar protein is shown in Fig. 14.5. Also, the added water in the form of liquid or ice acts

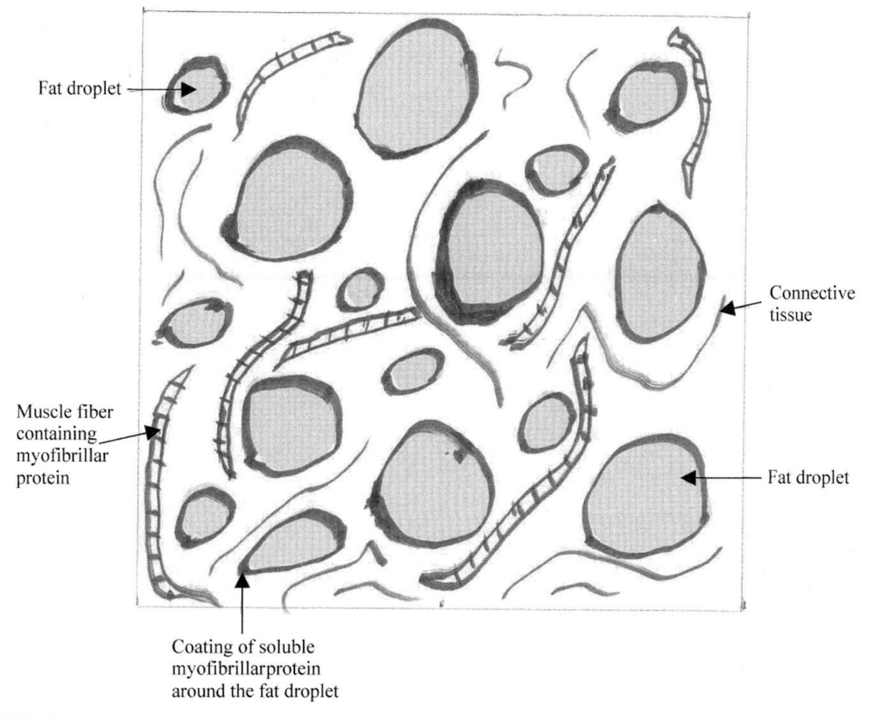

FIG. 14.5

An example of the physical structure of an emulsion matrix for a batter-type sausage.

to bring about proper texture, prevents excessive heating during chopping, and dissolves the curing ingredients during the batter-emulsion matrix formation process.

BATTER-TYPE SAUSAGE MANUFACTURE

GRINDING OF RAW MEAT MATERIALS (FOR COOKED AND SMOKED BATTER SAUSAGES)

The raw meat materials can be coarse ground (3/4–1 in. grinder plate) or fine ground (1/8–1/4 in. grinder plate). Coarse-ground meats are transferred to a bowl chopper (silent cutter), whereas fine ground meats are placed into an emulsion mill.

CHOPPING FOR COOKED AND SMOKED BATTER–EMULSION-TYPE SAUSAGE MANUFACTURE

Bowl chopper

Chopping ground meat in a silent chopper or bowl chopper (see Fig. 14.6) starts the process of developing a meat or sausage emulsion matrix for the batter preparation. Lean ground meat is added to the bowl chopper along with one half of the water or ice used in the formula, and salt, curing ingredients, phosphates, and seasonings, such as white pepper, paprika, nutmeg, coriander, ginger, and garlic, are added systematically into the chopper and finely chopped for several minutes. Once the mixture reaches 36–40°F, the fat portion, remaining water, and other ingredients

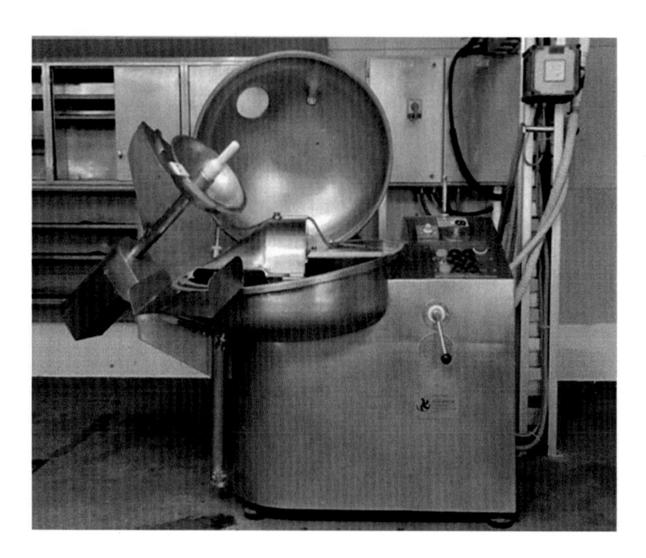

FIG. 14.6

An example of a silent chopper used to prepare emulsions for sausage manufacture.

Courtesy, Iowa State University Meat Science Laboratory.

are added and rapidly comminuted for several minutes until a stable batter-emulsion matrix (Fig. 14.5) is produced. Binding of water is especially critical in the production of low-fat hot dogs. Nonmeat proteins, such as soy, are often used in the low-fat hot dogs.

Smaller fat droplets are easier for proteins to stabilize. Chopping the batter to the temperature that fat droplets may start to melt helps to decrease the size of the fat droplets. Therefore final chopping temperature is critical to establish a stable emulsion matrix for the batter, and the final chopping temperature varies with the type of meat used in the formula for the sausage that is chopped. For example, the temperature for final chopping of pork is 55–60°F, poultry is 34°F, and 65–70°F for beef. In general, more saturated fats in beef tissue require a warmer temperature for melting and stabilization of the lipid by the fat-binding proteins. The amount of fat must be held within certain limits if the sausage is to maintain a proper and stable emulsion matrix structure during processing and handling. There must be enough soluble protein available so it can surround the fat droplets and rapidly coagulate during the heating process. This prevents the escape of the liquid fat from the fat particles that are surrounded by the meat protein and helps to form a stable emulsion matrix. Proteins and fat have an attractive interface that act as a glue to hold together the emulsion matrix for a stable product. An example of the proteins that surround the fat particles is shown in Fig. 14.5.

Emulsion mill and preblending of the batter

The emulsion mill is used to produce a batter for sausage products that have a fine texture, and the meat used is ground through a fine (1/8 in.) grinder plate (see Fig. 13.2 in Chapter 13). When preblending methods are used in the manufacture of the batter for specific sausages, the ingredients are placed through the emulsion mill 24 h before the scheduled manufacture of the sausage, and it is stored under refrigerated conditions. The procedure starts with the grinding of the lean meat followed by the grinding of the fatter meat. The next step is to mix the lean meat with salt, nitrite, and phosphate. The next procedure is to add the fatter meat and all other ingredients. At this stage, the temperature of the mixture should be checked and corrected to the temperature needed for the specific type of sausage. The next step is to run the sausage ingredients through the emulsion mill. The products should be chopped to the end point temperature for the specific sausage. The preblended product can enhance the sausage quality and make for more efficient use of labor and equipment.

Consumers are demanding lower fat products, and frankfurters are no exception. Higher fat (30% fat) wieners are still popular, but lower fat (15%–20%) to fat-free wieners have become more popular. The manufacture of lower fat wieners and fat-free wieners involves the replacement of fat with water in the wiener formulation. This formulation change can present a problem for developing a strong emulsion matrix for the low-fat wiener. To correct this problem, nonmeat ingredients are added. Hydrocolloids such as carrageenan and alginates, oat bran and oat fibers, milk products, serum proteins, plant proteins such as soybean products, yeast hydrolyzates, hydrolyzed vegetable proteins, and phosphates are examples of products used to prepare a stable emulsion matrix for low-fat wieners when an emulsion mill is used for the manufacturing process.

STUFFING AND LINKING FOR COOKED AND SMOKED EMULSION SAUSAGES

The chopped sausage is typically transferred to a type of continuous vacuum stuffer. Frankfurters and similar products are then stuffed into cellulose casings or natural sheep or hog casings or edible collagen casings. Frankfurters, when stuffed into small-diameter cellulose casings, are ultimately merchandized as skinless hot dogs. During heat processing, a thin coating of coagulated protein surrounds the exterior of the hot dog. The skin that is formed at the surface is important because it allows for efficient peeling of the cellulose casings when producing skinless franks. Continuous frankfurter processing lines are capable of producing more than 12,000 pounds of frankfurters per hour (Fig. 14.7).

For stuffing bologna for luncheon meats, 3.5- to 4-in. diameter fibrous casings are used, and for encasing ring bologna, 1.25- to 1.75-in. diameter is used. All casings should be stuffed to capacity to prevent air pockets. By using vacuum stuffers, air pockets can be eliminated by obtaining the maximum fill of the casing and oxygen can be minimized to prevent oxidative rancidity. Most sausages are uniformly linked in natural, collagen, and cellulose casings by using large, industrial-type linkers (Fig. 14.7). Some small meat plants and home operations may still link the sausage by hand into natural and collagen-type casings.

FIG. 14.7

An example of a sausage processing unit.

Courtesy, Marel Meat Processing Inc., Des Moines, Iowa.

SMOKING FOR COOKED AND SMOKED SAUSAGES

Natural wood smoke or liquid smoke are most effectively applied on the surface of sausage when conditions in the smokehouse are slightly moist but not wet. If it is too moist, there will be streaking, and if it is too dry, a limited amount of smoke will adhere to the sausage surface. Length of time to smoke the product depends on smoke density and air velocity in the thermal processing unit or smokehouse (see Fig. 13.21). After the smoking and cooking process is completed, the sausage is chilled, and if the sausage has cellulose casings, they are peeled to remove the cellulose casings with a high-speed peeling machine.

EXAMPLES OF SMOKED AND COOKED BATTER SAUSAGES

Of the cooked and smoked batter sausages, frankfurters (Fig. 14.8; sometimes called franks, wieners, or hot dogs) are the most popular of all the sausage products produced in the United States. They represent more than 25% of all the sausage products sold in the United States. A typical frankfurter will have a composition of 60% beef and 40% pork. Wieners can also be made of 100% beef, 100% pork, 100% poultry meat, or a combination of these meat sources. The wieners can vary in size and style for different markets. The largest diameter is the frankfurter, and the smallest diameter is the Vienna-style wieners (Fig. 14.9). The Vienna-style wiener takes its name from the city of Vienna, Austria. Another style of Vienna sausages is marketed in a can, and it is stable at room temperature (Fig. 14.10). Large bologna is also a very popular sausage in the United States, and it accounts for about 20% of the sausage consumption. Its processing formulation is similar to franks, but the bologna casing is much larger in diameter than wieners.

Some examples of well-known nonbatter smoked and cooked sausages are smoked pork sausage, polish sausage (Fig. 14.11), and smoked thuringer. These products are

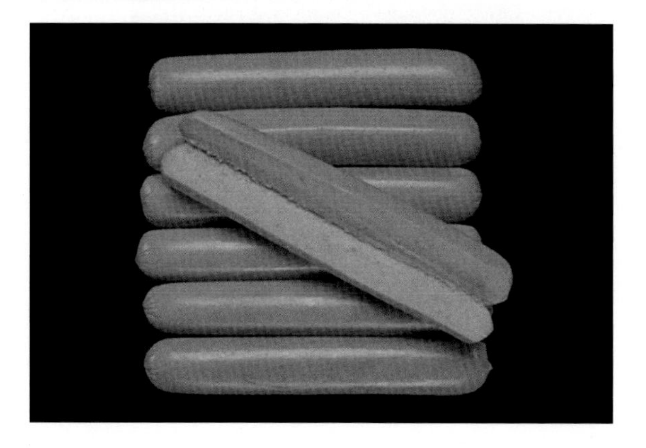

FIG. 14.8

An example of frankfurters (wieners and hot dogs) as a cooked and smoked sausage.

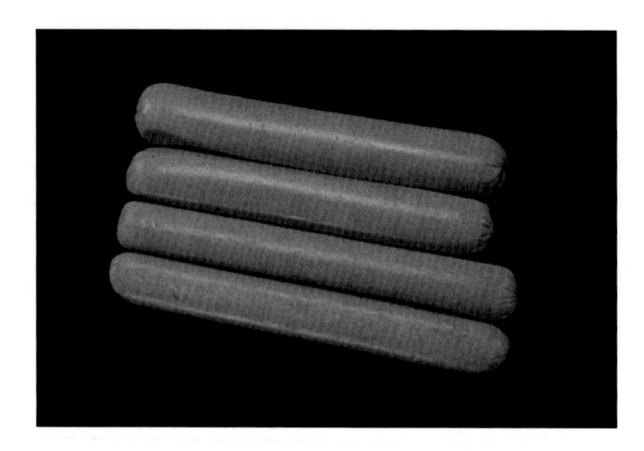

FIG. 14.9

An example of Vienna-style wieners. Vienna-style wieners are typically smaller in diameter and longer than the traditional frankfurter. Two wieners on the bottom of the figure are the Vienna-style wieners.

FIG. 14.10

An example of Vienna-style wieners that are canned and often consumed on camping trips.

FIG. 14.11

An example of a coarse-ground polish sausage.

coarser in texture than frankfurters and do not require a bowl chopper or emulsion mill for their production. The product is ground for a particle size of 1/4 to 1/8 in.

Braunschweiger (Fig. 14.12) is an example of cooked sausage that is made from liver, pork trimmings, jowls, bacon ends, and onions. The livers are ground and then placed in a chopper with salt and cure ingredients. Then ground pork, jowls, bacon ends, and other seasonings such as white pepper, dextrose, nutmeg, ginger, allspice, cloves, and water are added to the liver mixture, and the formulation is chopped to approximately 65°F. The processed product is encased in fibrous casings and either cooked in water at 180°F or transferred to the thermal processing unit and steam cooked to 155°F internally. After cooking, Braunschweiger must be refrigerated and stored under refrigerated conditions to preserve flavor and prevent microbial growth.

FIG. 14.12

An example of Braunschweiger or liver sausage.

Courtesy of Smithfield Foods.

SEMIDRY SAUSAGE

Semidry sausages are unique sausages, and most semidry sausages depend on bacterial fermentation for lactic acid production and preservation. Some examples of semidry sausages are summer sausage (Fig. 14.13), snack sticks (Fig. 14.14), thuringer, and cervelats. The terms "cervelat" and "thuringer" are sometimes used interchangeably with summer sausage, and there are many styles of these sausages.

MANUFACTURE OF SEMIDRY SAUSAGE

The manufacture of semidry sausage starts with the grinding and chopping of the meat ingredients, usually pork and beef. The chopped meat is mixed with the cure ingredients, the seasonings, starter culture, and dextrose (sugar) for lactic acid production. After mixing, the sausage is stuffed into a casing, fermented, cooked, and smoked. As the name (semidry) indicates, they are not completely dry like dry sausages. The semidry sausage is usually fermented for about 8–12 h to a pH of 4.8–5.2

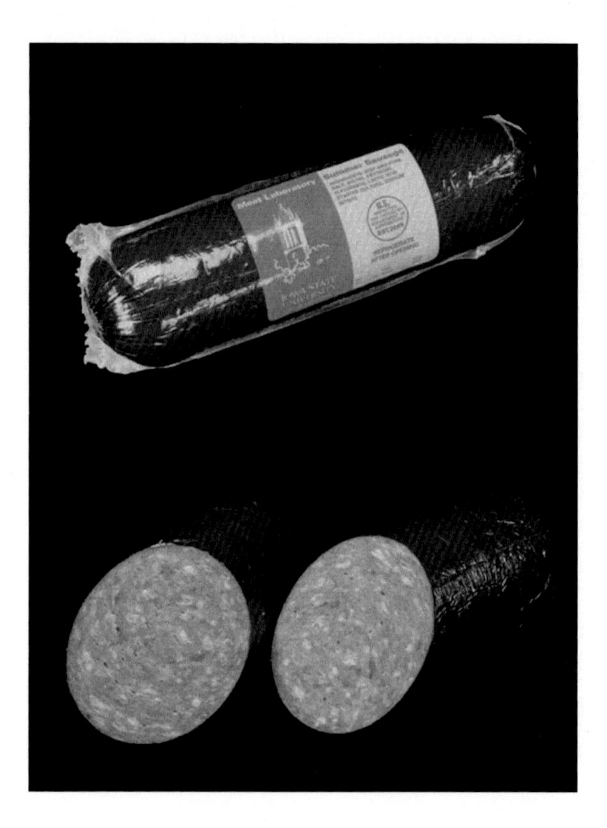

FIG. 14.13

An example of summer sausage as a semidry sausage.

FIG. 14.14

An example of semidry sausages in the form of snack sticks.

at 80–110°F at a humidity of 95%. It is then smoked and cooked to an internal temperature of 140–155°F for most sausage varieties. The final cooked temperature may vary depending on the type of semidry sausage manufactured.

DRY SAUSAGES

Dry sausages are one of the oldest forms of preserved meat. The preservation results from salting, fermentation for lactic acid production, and drying of the sausage product. Although drying and salting are of importance to the preservation of dry sausages, an important factor for sausage preservation is the development of lactic acid by bacterial fermentation. An example would be the fermentation of added dextrose to lactic acid resulting in a decrease in the pH of the sausage mixture from 6.0 to 5.6 to 5.0 or lower. It is a good policy to use meat for the dry sausage manufacture that has a pH lower than 6.0. A high pH meat (pH above 6.0) may interfere or prevent the required drop in pH in the sausage mixture during the manufacturing process. When appropriate acid conditions are obtained in the dried sausage, microbial growth will be stopped by heating, and the drying process is started. The low pH (high lactic acid content) of the sausage also provides the characteristic "tangy" flavor to the dried sausage. To be classified a dry sausage, the sausage must have low moisture content (30%–40%, Aw less than 0.91) or a moisture-protein ratio (M:P) of 2.3 to 1.0 or less. For example, Genoa salami has an M:P ratio of 2.3 or less, hard salami has an M:P ratio of 1.9 or less, and pepperoni 1.6 or less. The combination of the low pH and the low water content makes conditions in the sausage favorable to prevent bacterial growth and survival. Under these conditions, the dry sausage can be stored at normal room temperature and still provide a safe product for the consumer. Examples of dry sausages are capicola, mortadella, and pepperoni (Fig. 14.15). Pepperoni is the most popular pizza topping in the United States.

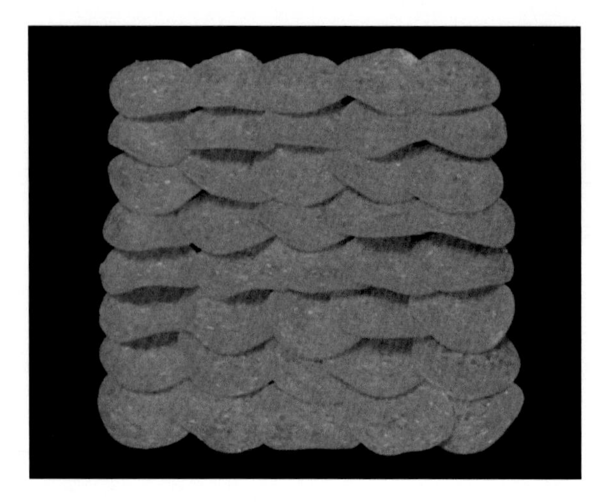

FIG. 14.15

An example of pepperoni as a dry sausage.

MANUFACTURE OF THE DRY SAUSAGE

The appropriate meat for dry sausage is ground or chopped (1/8 in. grinder plate) at 34–36°F to ensure that the fat and lean particle size is correct and that fat smearing on the sausage surface does not occur. The grinding temperature is critical for the production of quality dry sausage. After the meat is appropriately ground, it is briefly mixed (2–3 min) with the cure ingredients, seasonings, dextrose (corn sugar), and the lactic acid-producing bacteria (starter culture). After these ingredients are blended with the meat, the sausage mixture, for example, hard salami, is stuffed into either natural casings or fibrous casings (2–3 in. in diameter).

Proper fermentation for lactic acid production requires 24–48 h at 75–80°F. After the fermentation is completed, the sausage is smoked and subsequently air dried. Another important feature for the manufacture of dry sausage is the required control of the air drying process. Depending on the type of dry sausage, the drying process could take several months. Special drying rooms are used, and the temperature can range from 45 to 55°F with a relative humidity of 75%–80%. The air exchanges in the drying rooms occur 15–25 times per hour. Dry sausages lose approximately 30% of their original weight in the drying rooms. An example would be hard salami (Fig. 14.16).

Good quality control of the drying process is essential for the production of quality dry sausage. If the product dries too rapidly, a case hardening (hard dry ring on the surface of sausage) can occur, and this can prevent moisture loss of the sausage and result in internal spoilage by anaerobic bacteria. Conversely, if the sausage dries too slowly in the presence of higher humidity, there may be excessive mold, yeast, and bacteria on the sausage surface, slime can develop, and the sausage will develop spoilage conditions resulting in off-flavors.

FIG. 14.16

An example of salami as a hard, dry-type sausage.

Courtesy of Smithfield Foods.

LOAVES AND LUNCHEON MEATS

Luncheon meat and loaves are usually made from fresh meat. They are fully cooked or baked. Some loaves are smoked and some are not smoked. An example of a non-smoked loaf is head cheese shown in Fig. 14.17.

FIG. 14.17

An example of a meat loaf used for sliced luncheon meat. A head cheese is shown previously.

Courtesy of Smithfield Foods.

Luncheon meats are often sliced and vacuum packaged at the processing plant and then transported to retail outlets. Luncheon meat from the loaves is ready to eat and should be eaten before 7 days after opening the package to obtain the best tasting product. Dutch ham; jellied tongue; old fashioned olive (Fig. 14.4), pickle, and pimento loaves; head cheese; and scrapple are some examples of luncheon meats and loaves. Each of these products has a distinctive appearance, flavor, and appetite appeal. Moreover, this class of sausage offers unique possibilities of producing luncheon meats with fruit and vegetable components.

LOAF MANUFACTURE

Seasonings and type of meat are important factors in the manufacture of cooked loaves. For specific loaves, there must be 65% meat, they may have 3% added water, and they may contain extenders and binders. The nonspecific loaves have no restrictions for formulations. Emulsions such as those used for frankfurters and bologna can be used for loaves, or raw meats can also be used for loaf manufacture. Also, raw meats can be ground to different sizes to obtain a variety of texture and mouth feel of loaves and luncheon meats. When pickles, pimentos, and olives are used for the loaves, they are the last to be added to the formulation and must be carefully mixed before cooking. Loaf emulsions are stuffed into casings or molds or pans for cooking (baking). Complete cooking is accomplished when the internal temperature reaches 155°F or higher. After cooking and cooling, the loaves can be sold whole for sale at delicatessens, or they can automatically be sliced into portion-controlled luncheon meat packages for eventual distribution and sale. Some examples are olive loaf, pimento loaf, head cheese (Fig. 14.17), and scrapple.

CASINGS

Casings can function as a packaging material when the ground and chopped or emulsified meat products are stuffed in a natural or synthetic casing. These casings act as packages or containers to form and hold the product in its desired shape and size, protect it from bacterial contamination, and offer an eye appealing entity for retail merchandizing.

NATURAL CASINGS

Natural casings (Table 14.2) were the original casings used to package sausage products. This was the natural way because there were no synthetic or manufactured casings available when sausages were produced centuries ago. Wurstmachers predominately used gastrointestinal tracts from animals slaughtered in the abattoir for the natural casings. In addition to the casings obtained from the gastrointestinal tract, other natural casings were obtained from the stomach and bladders of beef, swine, and sheep. Some casings are made from the lining of the esophagus (termed "weasands") of beef. Natural casings are collected at the time of slaughter, flushed with

Table 14.2 Examples of natural and synthetic casings used for various kinds of sausage products

Casing	Type of casing	Packaged product
Natural casings		
Pork casings	Hog bungs	Braunschweiger
	Middles (middle portion of large intestine)	Dry sausage
	Small casings (small intestine)	Frankfurters
	Stomachs	Head cheese
	Bladders	Minced luncheon meat
Beef casings	Middles (large intestine)	Cervelat, salami
	Rounds (small intestine)	Bologna
	Bungs (trade name cecum)	Berliner, cappicoli
	Weasands (lining of esophagus)	Long bologna
	Bladders	Luncheon meat
Sheep casings	Middle (large intestine)	Frankfurters
	Small (small intestine)	Frankfurters
Synthetic casings	Cellulose	Frankfurters
	Plastic	Braunschweiger
	Fibrous	Summer sausages, bologna

water, packaged, and purchased by casing processors. They are thoroughly cleaned, graded, salted, and packaged for purchase and distributed to meat processors. They will be stored under refrigeration to prevent spoilage. The small casings (small intestines) from sheep are used to package frankfurters. Small casings from sheep are recognized especially for their ability to impart "old world" appearance and bite in hot dogs. The small intestine casings from pigs are typically used for sausages such as bratwurst. Beef middles (large intestines) are for packaging salami. The natural casings (hog bungs) are the straight sections of the large intestines and are used to encase, for example, Braunschweiger.

The advantage of using natural casings today in certain sausages is to meet the demand for traditional "old world" appearance. These qualities have a certain appeal to traditional tastes and ethnic groups. Another benefit of natural casings is their property to change permeability during cooking and smoking (from permeable to impermeable) and to have excellent elasticity and tensile strength.

MANUFACTURED CASINGS

Only small quantities of natural casings are used today because man-made or manufactured casings have the advantages of lower costs, more convenience, less microbial contamination, and greater safety. Synthetic casings (Table 14.2) are made from plastic, cellulose, and collagen materials and by coextrusion. These account for most of the casings used today to stuff sausage products.

There are three kinds of cellulose casings: small, large, and fibrous. Small casings give frankfurters the desired or required uniformity in diameter, length, and stability to withstand processing temperatures. They can be either clear or colored. Small cellulose casings are used in the manufacture of skinless frankfurters. On the other hand, large fibrous casings are used for encasing large bologna and cooked salami. Cellulose and fibrous casings are inedible and should be removed before slicing or eating.

Portion-controlled luncheon meat specialties, summer sausage, and large bologna are round products requiring the structural strength of reinforced cellulose or fibrous casings. As a result, uniform diameter slices are obtained in automated slicing and packaging lines. They are also used to package square sausage products as well as other smoked and processed sausages.

Collagen casings are an edible replacement for natural casings. Collagen casings are available in both small and large sizes. Collagen is a connective tissue protein that is found in animals. The hide or skin has a collagen layer that is used in the manufacture of the collagen casings.

Impermeable plastic casings such as Saran and Cryovac wraps are used for packaging Braunschweiger because they are suitable for water cooking after stuffing. These casings have the properties of low moisture vapor and low oxygen transmission to prevent weight loss.

Proper packing of sausage products for retail sales consists of an attractive, informative package that has an extended shelf life. Packaging systems and their specific characteristics will be discussed in detail in Chapter 15.

SUMMARY

Sausage processing and production companies represent a very large industry in the United States and countries around the world, particularly in Europe. Sausage is one of the oldest forms of processed food going back 1500 BC in ancient China. The raw materials for sausage manufacture are derived from skeletal muscle as well as variety meats such as heart and liver. Also, fat tissue is a major raw material for sausage manufacture. Nonmeat ingredients are major contributors to the quality, taste, and flavor of sausage. Water, salt, nitrites, sugars, ascorbates, and phosphates are critical nonmeat ingredients. The importance of these ingredients is discussed in the chapter. Other important nonmeat ingredients are spices, flavorings, and antioxidants. These compounds have significant influences on the flavor of the sausage. Some examples of fresh sausage are bratwurst and country-style pork sausage. Cooked and smoked sausages made from fresh meat are very popular, and this classification of sausages includes frankfurters or wieners, polish sausage, and bologna. The characteristics and production of these sausages are discussed in the chapter. After the sausage formulation is ground, chopped, seasoned, and mixed, the sausage material is stuffed in a casing and then it is cooked and smoked to add flavor and develop the cured meat color. Dry and semidry sausages are also very popular. Some examples are pepperoni, cappicoli, and summer sausage. Luncheon meats are another style of sausage production. Some luncheon meats contain olives or pimentos and are referred to as olive or pimentos loaves. Scientific information and concepts are presented for each of the sausage processing topics.

QUESTIONS FOR STUDY AND DISCUSSION TOPICS

1. Define the creative foods called sausage.
2. Provide a brief history of the sausage industry.
3. Describe the type of meat used for sausage manufacture.
4. What does fat tissue contribute to the quality of sausage products?
5. List five nonmeat ingredients used in the manufacture of sausage, and briefly describe their role in the sausage production process.
6. Describe two types of fresh sausage.
7. Describe three types of cooked and smoked sausage.
8. What are the ingredients for the production of Braunschweiger, and how is it processed into sausage?
9. What are the functions of casings used to stuff sausage products such as frankfurters?
10. What is the function of a thermal processor for sausage production?

CHAPTER

Packaging for meat and meat products

15

INTRODUCTION

Selection of appropriate packaging is a critical and essential consideration for meat and meat products that, if done correctly, will enable products to be stored and distributed over extended periods of time without losing quality or compromising safety. Generally speaking, the purpose of packaging is to protect products from the environment in order to prevent physical contamination (dirt, dust, bacteria, etc.) and from the chemical changes that result from environmental conditions (light, oxygen, evaporation of water, heat, etc.).

In addition to protection, packaging provides a very important means of communicating with consumers by including information about the product (ingredients, nutritional information, net weight included, handling instructions), the processor (name and address), and useful information for preparation and serving. A package can also provide an opportunity for consumers to view the product, either through a transparent film package, a transparent film window, or a printed picture on the package. For example, Fig. 15.1 is a package that displays the product for consumer viewing and provides information on the label. Additional information on nutrition, ingredients, and handling instructions are provided on the back panel. Packaging can also communicate information about the processes involved with production and processing of the product. Examples include "organic," "natural," "free range," "non-gmo," "reduced fat," "gluten-free," and other production practices that are becoming increasingly important to consumers who seek assurance of certain quality expectations or perceptions associated with those practices.

Finally, effective packaging must be easy and convenient for consumers to use. This includes easy-to-open packages with reclosure options (Fig. 15.2), packages that provide the desired quantity of product (single serving, family size, etc.), and a size and shape that facilitates easy storage (stackable for example) and handling in the home (Fig. 15.3).

Packaging becomes a very effective marketing and sales tool if it provides all of the features that consumers expect. This is becoming more and more important as the food industry becomes increasingly competitive and as consumers become less and less familiar with food production and processing practices, but have greater and greater expectations about the food that they buy.

The Science of Animal Growth and Meat Technology. https://doi.org/10.1016/B978-0-12-815277-5.00015-9

FIG. 15.1

Vacuum package that provides product viewing and label information.

Courtesy of Bemis Company, Inc.

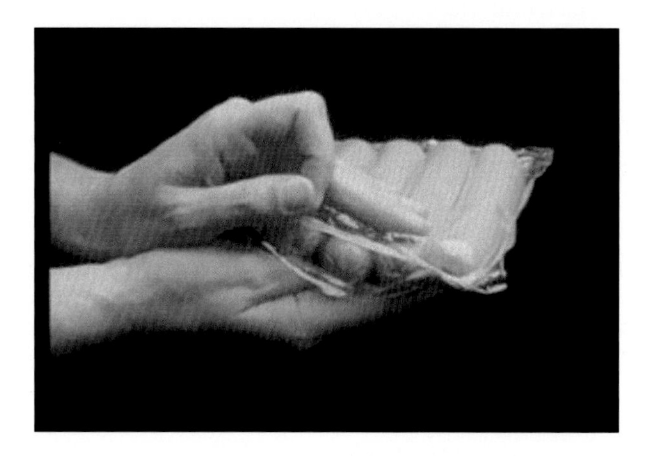

FIG. 15.2

An easy-to-open package for consumer convenience.

Courtesy of Bemis Company, Inc., Dr. Dan Siegel and American Meat Science Association.

PACKAGING REQUIREMENTS FOR MEAT AND MEAT PRODUCTS

FRESH MEAT

The requirements for effective packaging of meat and meat products are very different when considering fresh, raw meat such as ground beef, steaks or chops versus processed products such as frankfurters, hams, and roast beef. Fresh meat is biologically active and contains many active enzymes and compounds that support

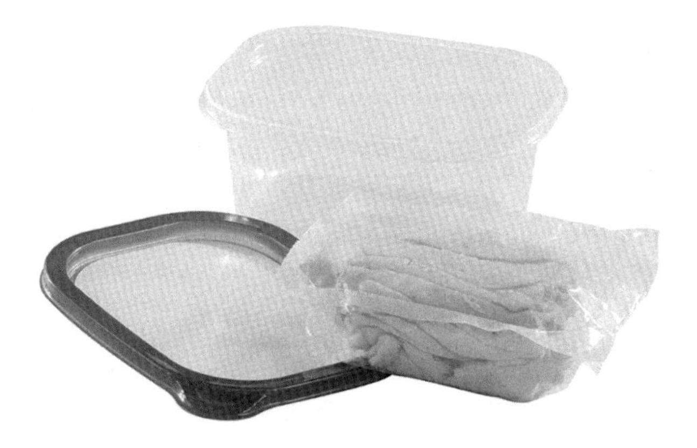

FIG. 15.3

Resealable packaging with stackable feature.

Courtesy of Bemis Company, Inc.

metabolic and respiratory activities that affect both the product and the environment around it. For example, fresh meat continues to consume oxygen and emit carbon dioxide, not unlike living muscle, albeit at a much slower pace. The result in a closed package is reduction of oxygen content and increase in carbon dioxide, with implications for both product color and product shelf life. Because color of fresh meat is critical to consumers' buying decisions, a major priority for fresh meat packaging is development and maintenance of desirable cherry red color for as long as possible.

Fresh meat will also contain a certain amount of unavoidable bacteria that grow readily on raw meat and eventually reduce product quality. Thus a second priority for fresh meat packaging is introducing technology that will slow bacterial growth and reduce product spoilage during storage and distribution. Because oxygen exposure provides the most common means of developing the cherry red color of fresh meat but also encourages rapid growth of aerobic spoilage bacteria, packaging technology for fresh meat has developed several alternatives to achieving both color and shelf life objectives.

PROCESSED MEAT

For processed meat products which include nitrite-cured products and uncured, cooked products, color is developed as part of the curing and/or cooking process, and most of the biological and microbiological activities are eliminated by the cooking process which inactivates both enzymes and microorganisms. As a result, product color is much more stable than for fresh meat and the potential shelf life is considerably longer. A major priority for packaging in this case is to eliminate or greatly reduce oxygen contact with the product to maintain color stability, flavor stability, and suppression of microorganism growth for as long as possible.

PACKAGING MATERIALS

Over the past 50 years or so, the meat industry has moved from butcher paper and simple film overwraps for retail sales to what is now primarily "case-ready" packaging with multilayer flexible films. Fig. 15.4 is an example of a packaging line for individually packaged steaks in trays that will be ready for retail display, and Fig. 15.5 is an example of case-ready packaging in flexible films. This means that a large share of meat products is packaged in centralized facilities, then transported to warehouses and retail stores in a form that can be simply placed on retail store shelves. This has been made possible by the development of complex flexible packaging films that provide strength, specific gas and water vapor barrier properties, printability, and good heat sealing properties. Because no single film can provide all of the necessary properties, most packaging films are typically laminates or coextrusions of multiple film materials to create the best combination of properties for a specific product. The necessary film properties can be considered in three general categories: strength, barrier properties, and sealing ability.

FILM MATERIALS THAT PROVIDE STRENGTH

To provide a package with puncture resistance, tear resistance, and the ability to tolerate abrasion, handling, and, in some cases, thermal processing, there are three flexible film materials that are commonly used: nylon, polyester, or polypropylene.

FIG. 15.4

Case-ready packaging line for steaks in trays.

Courtesy of Sealed Air Food Care Corporation.

FIG. 15.5

Case-ready packaged steak in flexible film package.

Courtesy of Bemis Company, Inc.

Nylon is recognized for its tensile strength, flexibility, and thermal tolerance and will also provide a good oxygen barrier when dry. However, it is a poor moisture barrier and will absorb moisture. Consequently, it is typically used as a middle layer for multiple layer films with other films providing the needed barrier properties and heat sealing ability, particularly for vacuum packages and cook-in bags.

Polyester is also noted for strength, flexibility, and thermal stability, and also provides a good base for printing. Polyester is not a good barrier to either gases or water vapor and is used primarily for its strength in vacuum packages, cook-in bags, and products that may be heat sterilized.

Polypropylene, in oriented form, provides good tensile strength and temperature tolerance. This film is also a good water barrier which make polypropylene a good choice for boil-in-the-bag products or steam-processed packaged products.

FILM MATERIALS THAT PROVIDE BARRIER PROPERTIES

Barrier properties for multilayer films are most often achieved with polyvinyl chloride, polyvinylidene dichloride (also known as Saran), ethylene vinyl alcohol, and aluminum foil. Polyvinyl chloride is clear, glossy, and has good tensile strength, though not comparable to nylon or polyester. It seals easily with heat and, in the plasticized form, is impermeable to water vapor but is oxygen permeable which lends this film to overwrapping of fresh meat to allow good color development.

Polyvinylidene dichloride (Saran) is well known for its barrier properties to both oxygen and water vapor. It is heat sealable, accepts printing well, and is a common inclusion in multilayer films where a good barrier is needed such as thermoformed, semirigid packages and in modified atmosphere packaging.

Ethylene vinyl alcohol is a film highly impermeable to gases such as oxygen and carbon dioxide, but can lose some of its barrier properties when exposed to high humidity. Consequently, it is best used as a core layer with films that protect it from moisture. This film material also provides good strength, elasticity, glossiness, and heat stability. Its most common use is for packages where a good oxygen barrier is critical.

The ultimate in barrier materials for packaging is aluminum foil. It is not only an outstanding barrier to oxygen and water vapor but also provides protection from light. Aluminum foil is most often used for products that are susceptible to flavor changes induced by light and oxygen, such as dried products. Aluminum foil is also often used for flexible retort packages that are sterilized by heat. This material has very little tear resistance, cannot be heat sealed, and will lose barrier properties if pin holes are allowed to develop. Consequently, it is necessary to combine aluminum foil with a sealant film as well as a film that will provide puncture resistance, tear resistance, and tensile strength.

FILM MATERIALS THAT PROVIDE PACKAGE SEALING ABILITY

Materials that provide the heat sealing properties necessary for multilayer film packages include polyethylene, Surlyn, and polystyrene. Polyethylene is available in three different forms with the low density form providing the most effective heat sealing material. Polyethylene is a good barrier to water vapor but a poor barrier to oxygen. Because this material heat seals very well at relatively low sealing temperature, it is frequently combined with barrier films and high strength films in meat packaging applications. Polyethylene is also a relatively low cost material which makes it attractive for heat sealing applications.

Surlyn is polymerized ethylene modified with the addition of sodium or metal ions that provide crosslinking between the polymers to increase rigidity. Surlyn has a broad heat sealing temperature range and provides good adhesion to other materials including aluminum foil, making it an effective sealant for many different packaging applications. While commonly used in meat packaging, Surlyn is a relatively expensive film.

Polystyrene is a hard, transparent material that is most often formed into rigid trays or tray lids. Foamed polystyrene provides the foam material often used for trays as well. Polystyrene is not a good barrier to either oxygen or water vapor and combining polystyrene with barrier films is necessary when barrier properties are important.

PACKAGING OPTIONS FOR FRESH MEAT

FILM-WRAPPED TRAYS

The most common package that consumers experience with fresh meat is the film-wrapped tray where a highly oxygen-permeable film, usually polyvinyl chloride, is stretched over a polystyrene tray (Fig. 15.6) that includes a soaker pad on the bottom

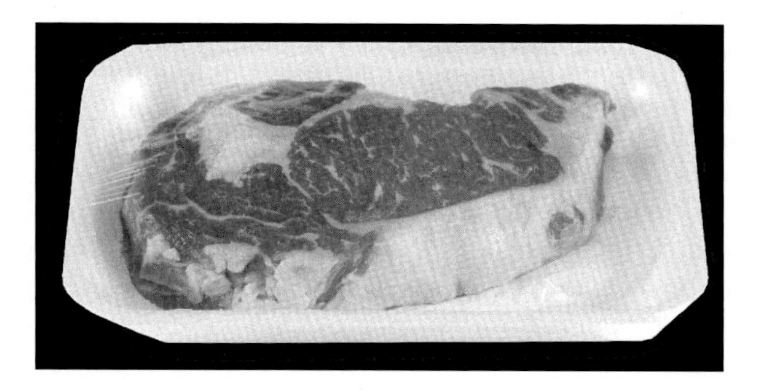

FIG. 15.6

Retail packaging utilizing oxygen-permeable film overwrapped tray.

Courtesy of Bemis Company, Inc.

of the tray to absorb purge. This film is clear, glossy, and allows atmospheric oxygen into the package to maintain the bright red color of bloomed fresh meat. However, because the meat is metabolizing the oxygen, oxygen concentration is slowly depleted and color fading typically occurs after about 3–7 days, depending on temperature and other environmental conditions.

MODIFIED ATMOSPHERE PACKAGING

To improve the color stability of fresh meat, modified atmosphere packaging (MAP) with either a high oxygen content or with carbon monoxide has been developed and is in common commercial use. For high-oxygen MAP, a gas mixture of 40%–80% oxygen is flushed into the package to replace the inherent air before the package is sealed. Typically, about 25% carbon dioxide is included because this gas slows bacterial growth. Carbon dioxide is not used at more than about 30% because the higher concentrations of this gas will cause meat discoloration. The remaining balance of the gas mixture is nitrogen, an inert filler gas. This high-oxygen MAP produces a deeper, longer lasting layer of oxymyoglobin, which is the pigment responsible for the cherry red color on the product surface. High-oxygen MAP improves color life of fresh meat to 10–14 days but comes with some potential concerns for flavor effects and reduced tenderness. Research has shown that the high oxygen concentrations can result in oxidized flavor and reduced proteolytic activity of enzymes. The enzymes responsible for meat tenderization during aging are oxidized at a faster rate by high oxygen environment which reduces the ability of these enzymes to break down meat proteins for improved tenderness. Another issue that has been raised concerning high-oxygen MAP is that the oxidation environment can reduce the thermal tolerance of the meat pigments. This can result in what is called "premature browning" in which the meat pigment denatures and turns brown during cooking at a lower than usual temperature. The implication is that the product might look as though it has reached an appropriate target temperature for bacterial safety when, in fact, it has not. This is not an issue if consumers use a thermometer to monitor cooking temperature but could be an issue for those that use visual appearance as an indicator of doneness. Because package interior gases need to be maintained at or near target concentrations, a high barrier film and a good sealing film such as ethylene vinyl alcohol is important for MAP to provide a good impermeable cover film combined with an impermeable base tray, or an impermeable pouch.

A low-oxygen alternative in MAP that is very effective for fresh meat is a gas-flushed atmosphere that includes 0.4% carbon monoxide with 25% carbon dioxide and the rest as nitrogen. Carbon monoxide at this concentration has been declared GRAS (generally recognized as safe) by the U.S. Food and Drug Administration (FDA) and approved for use for meat products by the U.S. Department of Agriculture-Food Safety Inspection Service (USDA-FSIS). At 0.4% in a gas mixture, carbon monoxide is not a safety issue, despite concerns raised by some consumer groups. Carbon monoxide forms a very strong bond with myoglobin that results in the same cherry red color as oxygen (Fig. 15.7). While there are very subtle differences in the color, the two pigments are visually indistinguishable. However, the color life of the carboxymyoglobin pigment is 28–35 days instead of 3–7 for oxymyoglobin in the case of atmospheric oxygen or 10–14 days in the case of high-oxygen MAP.

Another fresh meat packaging alternative that retains the attractive cherry red color is a novel packaging film developed by Bemis Corp. in 2010. This packaging system uses a film that contains a very low concentration of nitrite (a few parts per million) embedded in the film structure. The package, containing the product, is sealed under vacuum to form a vacuum package. When the film is in contact with the product surface, a small amount of nitrite is released to form nitric oxide which reacts with myoglobin to form nitric oxide myoglobin, a pigment that is cherry red and indistinguishable from oxymyoglobin and carboxymyoglobin (Fig. 15.8). As long as a vacuum is maintained, the red color is maintained as well. The use of vacuum, however, also means that shelf life, including color is stable for more than 35 days.

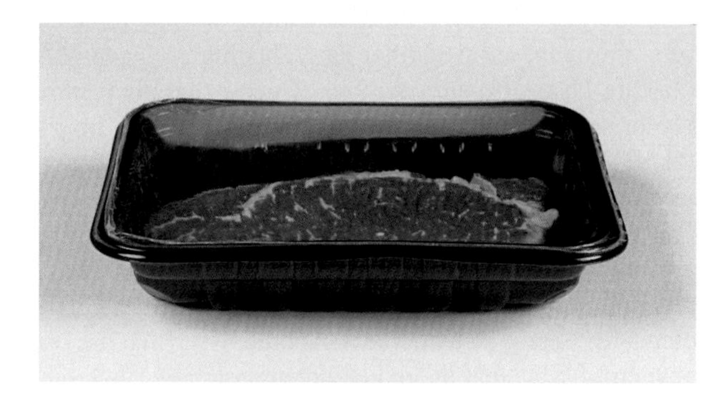

FIG. 15.7

Modified atmosphere package (MAP) utilizing 0.4% carbon monoxide in the gas mixture to develop cherry red meat color.

Courtesy of Bemis Company, Inc.

FIG. 15.8

Conventional vacuum package *(top)* and vacuum package in nitrite-embedded film *(bottom)*.

Courtesy of Bemis Company, Inc., Dr. Dan Siegel and American Meat Science Association.

The concentration of nitrite that is used in the film is too low to result in cured color when the package is opened or when the product is cooked.

CONVENTIONAL VACUUM PACKAGING

Conventional vacuum packaging with high-barrier films is not utilized for most retail packaging because this environment does not allow for oxygen-driven color bloom. However, conventional vacuum packaging is the system of choice for primal and subprimal cuts for wholesale distribution. A good vacuum package removes essentially all oxygen and consequently suppressed bacterial growth while preserving the ability of the meat to form bloomed red color when the package is opened and exposed to oxygen. Because the meat is metabolically active, any small amounts of residual oxygen that remain in the package after sealing will be metabolized and converted to carbon dioxide which helps to improve shelf life. Vacuum-packaged primals and subprimals have a potential shelf-life of up to 90–120 days. Because the color of vacuum-packaged fresh meat is a dark purplish red rather than cherry red (Fig. 15.9), the conventional vacuum system has never been well accepted by consumers in retail stores despite numerous efforts to market fresh meat in vacuum packages. Consequently, vacuum-packaged wholesale meat cuts are usually first

FIG. 15.9

Conventional vacuum package for subprimal and primal cuts showing typical purple-red color development.

Courtesy of Bemis Company, Inc.

opened, cut into smaller units, and repackaged in a packaging environment that achieves a cherry red color for retail display.

A novel approach to the use of vacuum followed by oxygen-permeable packaging for retail fresh meat was developed by the Sealed Air Corp. in which the product is packaged in both a barrier film and an oxygen permeable film. The meat is vacuum packaged in a multilayer film that has an outer barrier film that can be peeled away to expose the oxygen-permeable film. Products packaged this way can be cut into retail portions and packaged under vacuum to realize the shelf life advantages of vacuum packaging. At the retail level, the outer film is peeled off, allowing oxygen content to increase and development of cherry red bloom for retail display.

PACKAGING OPTIONS FOR PROCESSED MEAT PRODUCTS

VACUUM PACKAGING

Processed meat products include nitrite-cured meat products such as frankfurters and hams, and noncured cooked products such as roast beef and roasted turkey breast. Because a major concern for maintaining the quality of these products is preventing contact with oxygen, packaging in vacuum is a common practice for these as well as fresh and frozen poultry, and wholesale primal and subprimal cuts (Fig. 15.10). Because the processed meats are cooked products, the number of microorganisms is significantly reduced and the anaerobic environment of a vacuum package is conducive to an extended shelf life of well over 120 days if a low temperature is maintained.

FIG. 15.10

Vacuum packaging applications for fresh, frozen, cooked, and cured meat products.

Courtesy of Bemis Company, Inc., Dr. Dan Siegel and American Meat Science Association.

The films used for these packages will include high-barrier materials such as polyvinylidene dichloride or ethylene vinyl alcohol, and are often designed to be heat shrinkable to provide a very tight adherence to the product. This provides an attractive package with excellent product visibility and serves to reduce product purge as well.

A few products such as canned hams and some canned luncheon meats such as SPAM and Vienna sausages are still packaged under vacuum in metal or plastic cans. Cans provide the necessary barrier properties to water vapor and oxygen as well as the rigidity needed for the intense cooking processes used to achieve pasteurization or sterilization as in the case of room temperature stable canned meat.

Vacuum packaging is also used for cook-in-the-bag products such as roast beef, hams, and roasted turkey breasts that are displayed and sold by delicatessens where the products are sliced to order. These films must provide a good barrier for storage of the finished products as well as strength to withstand the cooking treatment. The barrier film material is usually ethylene vinyl alcohol and the base material for the interior of the package is a form of nylon or Surlyn. An unusual aspect of these films is that they also include a specifically designed inner layer or coating that is in direct contact with the product and that will physically bond to the product surface during cooking to eliminate purge. This interior surface material is highly specific for each type of product, even to the point of being slightly different for different species of product such as beef, pork, or turkey. The cook-in-the-bag process produces an extremely tight-fitting package that is attractive in delicatessen display.

MODIFIED ATMOSPHERE PACKAGING

While vacuum packaging is very effective for preserving color and inhibiting microbial spoilage of processed meats, the external physical pressure exerted on the product by the atmosphere is problematic for sliced products because slices or pieces will stick together as a result. In this case, a foam tray with a built-in barrier with a barrier film lid or a flexible pouch can be used (Fig. 15.11). Instead of using vacuum for these packages, headspace of the tray or the interior of the pouch is flushed with a mixture of 70% nitrogen and 30% carbon dioxide gases, or in some cases, 100% nitrogen. This achieves an oxygen-free environment without the physical pressure that is characteristic of a package under vacuum. Some dried products that are sensitive to light-induced deterioration such as jerky and dry sausage may be packaged in aluminum foil-laminated pouches to block both oxygen and light exposure.

FIG. 15.11

Gas-flushed package for processed meat products.

Courtesy of Bemis Company, Inc.

SPECIALIZED PACKAGING APPLICATIONS

There are several specialized packaging applications that provide improved control of shelf-life and microbial growth for specific types of products. These are categorized as "active packaging" or "intelligent packaging." Active packaging means that the package releases or absorbs compounds or gases during storage that change the package environment. Intelligent packaging includes a component that monitors conditions of the package in some fashion and provides an indication of the package history.

ACTIVE PACKAGING

Active packaging is a practice that is common for some types of foods. It can include oxygen scavengers, moisture absorbers, carbon dioxide emitters, sources of antimicrobials or antioxidants that are slowly released, and ultraviolet light barriers. Oxygen scavengers are probably the most common and are used for products sensitive to small amounts of residual oxygen in the package. Oxygen absorbers consist of a small sachet containing iron powder that is inserted into the package with the product and are most useful for gas-flushed or modified atmosphere packages where there is a headspace (Fig. 15.12). The iron is readily oxidized by oxygen and the result is removal of the residual oxygen in the package. Because it is virtually impossible to remove all of the oxygen from a package with gas flushing, the oxygen scavenger can play an important role in the product shelf life. Some film manufacturers have developed packaging materials that incorporate an oxygen absorber into the film structure which can then be used for the interior layer of a multilayer film package.

Moisture absorbers are usually silica gel sachets that can be used to maintain a low humidity environment where that may be needed and are used for some consumer products such as electronics. However, the use of moisture absorbers is not common for products such as processed meats that have a high moisture content. The soaker pads used in trays of fresh meat to absorb purge can be considered another form of moisture absorber. These are common on the bottom of fresh meat trays and prevent the unattractive appearance of free purge in the package.

FIG. 15.12

Gas-flushed master package for distribution of retail-ready packages. Note the oxygen scavenger sachet in the master package for absorbing any residual oxygen that might remain after gas flushing.

Courtesy of Bemis Company, Inc.

Carbon dioxide emitters provide a means to increase the carbon dioxide content in a package for greater antimicrobial impact. While not as common as oxygen scavengers, those used typically function by utilizing ascorbic acid and sodium bicarbonate to generate carbon dioxide and will also absorb oxygen in the process, thus providing a dual function.

Films that can include antimicrobial compounds or antioxidants that are released when in contact with the product have been a focus of many recent research and development efforts, particularly with the increased emphasis of food safety. A large number of antimicrobials have been studied including organic acids, bacteriocins, antibiotics, plant extracts, and essential oils. While many of these have been shown to be effective, commercial applications have been limited due to many factors, including regulatory approval of the specific antimicrobials and the added cost. However, this technology represents a great deal of potential for improved microbial control.

Antioxidants such as butylated hydroxyanisole (BHA) and butylated hydroxytoluene (BHT) are sometimes incorporated into packaging films to stabilize the film structure and prevent deterioration of the film. These antioxidants then have potential to migrate into the product and provide antioxidant protection. Natural antioxidants such as rosemary, tocopherol, and plant extracts have also been used for active antioxidant packaging systems.

Ultraviolet light is a strong catalyst for autooxidation of myoglobin to metmyoglobin which changes fresh meat color to an undesirable brown. The incorporation of ultraviolet light blockers into packaging films has been studied as a means to improve color stability of fresh meat. However, these films also block some of the visible light as well and the product becomes less visible as a result. Currently, the best approach to concerns about exposure to ultraviolet light and subsequent oxidation remains an aluminum foil package combined with elimination of residual oxygen.

INTELLIGENT PACKAGING

Intelligent packaging involves a packaging system that includes a device or compound that monitors internal or external package conditions and provided information or indication about the history of the package or product. For example, indicators have been developed that can be attached to a package surface and that will change color when certain cumulative temperature limits have been reached during storage or distribution. These serve to communicate incidents of temperature abuse if that should occur. Several types of indicators have been developed to monitor and show things like oxygen concentration, microbial metabolites as indicators of spoilage, bacterial pathogen growth, storage time, and humidity, all of which are intended to provide information on the real-time status of product quality. While several such indicators have been developed, they have not found widespread use in the meat industry due to the added cost of the indicators.

SUMMARY

Packaging provides protection for meat and meat products from environmental contaminants during storage and distribution, creates a suitable environment to maintaining product quality, can be used to achieve greatly improved shelf life, and provides information to consumers about the product contained within a specific package. Meat and meat products fall into two general categories for packaging requirements. Fresh meat, for example, must develop and retain an attractive red color for retail sales, and this can be accomplished with oxygen-permeable films, modified atmosphere packages containing high oxygen concentrations or carbon monoxide gas, or with vacuum packages using a nitrite-containing packaging film. Processed meat products, including nitrite-cured products and noncured cooked products require packaging that excludes oxygen, thus vacuum packaging in high-barrier packages is common, though gas-flush and modified atmosphere packaging that excludes oxygen are also effective and are in use for some products.

Modifications in film materials and new technologies have been developed to offer active packaging systems that modify package conditions during storage for greater effectiveness, and intelligent packaging systems that monitor and provide information about the environment in and around the package over time. These systems continue to be studied and refined and are likely to play a bigger role in the future.

QUESTIONS FOR STUDY AND DISCUSSION TOPICS

1. Why do fresh meat and processed meat products require very different packaging system?
2. What is the most important consideration for effective packaging (a) of fresh meat products, and (b) of processed meat products?
3. What are the basic requirements of the film materials to be used for flexible film packaging of fresh and processed meats?
4. What are the advantages and potential disadvantages of using a high oxygen concentration for modified atmosphere packaging of fresh meat?
5. What is the role of carbon monoxide meat packaging?
6. What is the role of nitrite in packaging of fresh meat?
7. Explain "modified atmosphere packaging," "active packaging," and "intelligent packaging."

Index

Note: Page numbers followed by *f* indicate figures and *t* indicate tables.

L

M

T

U

Printed in the United States
By Bookmasters